Laser Radar Systems

For a complete listing of the *Artech House Radar Library*,
turn to the back of this book . . .

Laser Radar Systems

Albert V. Jelalian

Artech House
Boston • London

Library of Congress Cataloging-in-Publication Data

Jelalian, A.V. (Albert V.)
   Laser radar systems / A.V. Jelalian.
     p.   cm.
   Includes bibliographical references and index.
   ISBN 0-89006-554-3
   1. Optical radar. I. Title.
   TK6592.06J44 1991   91-36803
   621.3848-dc20   CIP

British Library Cataloguing-in-Publication Data

Jelalian, Albert V.
   Laser radar systems.
   I. Title
   621.3848

   ISBN 0-89006-554-3

© 1992 Al Jelalian

By Artech House
685 Canton Street
Norwood, MA 02062

All rights reserved. Printed and bound in the United States of America. No part of this book may be reproduced or utilized in any form or by any means, electronic or mechanic including photocopying, recording, or by any information storage and retrieval system, without permission in writing from the publisher.

International Standard Book Number: 0-89006-554-3
Library of Congress Catalog Card Number: 91-36803

10 9 8 7 6 5 4 3 2 1

*Dedicated to
my wife and children,
Mary Berjouhi,
Alan, and Leslie,
who make it all worthwhile;
my parents and brother
Siragan, Zarouhi, and Lincoln
who made it possible;
family, colleagues, and friends
who make it fun.*

Contents

Acknowledgments	xi
Chapter 1 Laser Radar System Theory	1
1.1 Electromagnetic Spectrum	1
1.2 Range Equation	3
1.2.1 Range Equation Dependence on Target Area	6
1.3 Detection Considerations	10
1.3.1 Noise Power Spectral Density	12
1.3.2 Receiver Detection Techniques	13
1.3.3 Background Noise Terms	13
1.3.4 SNR Expression Development	16
1.3.5 Receiver Noise and Signal Terms	17
1.3.6 Comparison of Incoherent and Coherent Detection Receivers	19
1.3.7 Noise Figure	22
1.4 Beamwidth	23
1.5 Search Field	26
1.6 Search Field Figure of Merit	27
1.7 System Optimization	29
1.7.1 Incoherent Receiver	29
1.7.2 Coherent Receiver	30
1.7.3 System Efficiency	30
1.8 Temporal and Spatial Coherence	33
1.8.1 Spatial Coherence	36
1.8.2 Spatial Coherence Alignment	36
1.8.3 Coherent System Alignment Requirements	36
1.8.4 Angular Scan Rate—Lag Angle Consideration	37
1.8.5 Effective Aperture Diameter for Coherent Receivers	38
1.8.6 Temporal Coherence	40

	1.8.7 Scan Spectral Broadening	41
1.9	Measurement Errors	42
1.10	Imaging Systems	48
	1.10.1 Angle-Angle Imaging	48
	1.10.2 Range-Doppler Imaging	49
	1.10.3 Range-Velocity Ambiguity	51
Chapter 2	Atmospheric Propagation	59
Chapter 3	Detection Probabilities and False Alarm Rates	77
3.1	Introduction	77
3.2	Target Detection Statistics (Microwave Radar Models)	78
3.3	Coherent Detection Laser Radar Models	88
	3.3.1 Introduction	88
	3.3.2 Atmospheric Turbulence Effects on Laser Radar Model	91
3.4	Probability of Detection and False Alarm Rates for Incoherent Detection Systems	97
	3.4.1 Introduction	97
	3.4.2 Incoherent Detection of Optically Noise (Poisson) Limited Receivers	100
	3.4.3 Incoherent Detection of Thermal (Gaussian) Noise-Limited Receivers	103
3.5	Mathematical Distinctions Between Microwave and Laser Photon Noise Limited Detection Statistics	109
	3.5.1 Introduction	109
	3.5.2 Swerling II	109
	3.5.3 Swerling IV	110
	3.5.4 Comparison of Swerling II and IV	110
	3.5.5 The Log-Normal Distribution	114
	3.5.6 The Rice Distribution	115
3.6	Lidar Signal Statistics in the Photon Count Limit	117
Chapter 4	Lasers	121
4.1	Introduction	121
4.2	Detectors	131
	4.2.1 Detector Responsivity	132
	4.2.2 Noise Equivalent Power	133
Chapter 5	Incoherent Receiver Detection Systems and Techniques	137
Chapter 6	Coherent Laser Systems and Techniques	151
6.1	Introduction	151
6.2	Fire Control Laser Radar Adjuncts	166
	6.2.1 Long-Range Laser Radars	169

Chapter 7 Atmospheric Laser Radar Systems		179
7.1	Introduction	179
7.2	Remote Atmospheric Measurement Techniques	183
	7.2.1 Incoherent Systems Approaches	184
	7.2.2 Incoherent Systems	184
	7.2.3 Coherent Systems	192
References		207
Appendix A		213
A.1	Introduction	213
A.2	Discussion of Reflectance Measurements	214
Glossary		279
Index		285

Acknowledgments

For over twenty-five years technologists in the laser radar field have developed lasers, optical components, and systems. Many scientific and engineering articles and reports have been written on the subject but few books have been written on the systems aspects of this technology.

In the winter of 1990, the Boston Chapter of the IEEE asked if I would deliver a lecture on laser radar systems to which I agreed. When they asked me what reference book I would be using, I was embarrassed to say that I was not aware of a suitable one.

This book was born at that point. The material within it is a collection of my notes, analysis, and reference material, which I hope will prove useful to those systems designers of the future, as well as those wishing to know more about the potential capability of these systems.

Over the years, I have had the good fortune to be associated with a group of scientific and engineering colleagues who have helped me with understanding this technology. To them, I owe a debt of gratitude. While all of them cannot be referenced, a few with whom I have traveled this road for a long time are—

Dr. Wayne Keene	Mr. Irving Goldstein
Dr. Knut Seeber	Mr. Ralph McManus
Dr. Charles Sonnenschein	Mr. Arthur Chabot
Dr. Gregory Osche	Mr. Clarke Harris
Dr. Martin Schilling	Mr. Alfred Legere
Mr. Harold Hart	Mr. Milton Huffaker
Dr. Hermann Statz	Mr. Edwin Weaver
Mr. Nathaniel Friedman	Dr. Herbert Grozinsky
Dr. Frank Horrigan	Mr. James Bilbro

A list like this is always incomplete. For those government and industry friends not listed, my apologies.

My appreciation also extends to Drs. George Dezenberg and Douglas Youmans for their reviews and comments, Ms. Margaret Smith, Mrs. Corinne Beach, Mr. Richard Lopez, and Mr. John Veracka for their efforts in converting my penmanship to this art form, and most especially to the Raytheon Company for allowing me the opportunity to contribute in this technical area.

Chapter 1
Laser Radar System Theory

1.1 ELECTROMAGNETIC SPECTRUM

Laser radar is a subject that can be viewed from different perspectives. Everyone looks at the electromagnetic spectrum a little differently. However, because the subject basically involves electromagnetic radiation, there are similarities that run through all observations because of the different wavebands. There are also significant differences.

The electromagnetic spectrum is divided into a variety of wavebands, as shown in Figure 1.1. Those involved with the initial nomenclature selection for the microwave wavelengths used many superlatives—*high, ultra-high,* and *super-high* frequencies. As a result of cryptic World War II efforts, letter nomenclature evolved for subdivisions of the microwave band, using such designations as *L-S-, C-, X-, K-,* and *Q-band*. Moving toward the shorter wavelengths, the *millimeter waveband* (W) is eventually reached. The other end of the electromagnetic spectrum, with few superlatives left, is named for where the human eye had response (the visible band from 0.4 to 0.7 μm). As the wavelengths became longer compared to 0.7 μm (red), the wavebands were successively called the *near-infrared,* the *mid-infrared* and *far-infrared* bands, as shown in Figure 1.1.

Useful lasers operate at wavelengths in the ultraviolet, visible, near-, mid-, and far-infrared regions, and can be used to transmit radiation to a target via a telescope. The scattered target radiation can be detected and processed in a laser radar receiver to remotely determine the target characteristics.

Laser radars constitute a direct extension of conventional radar techniques to very short wavelengths. Whether they are called *lidar* (light detection and ranging) or *ladar* (laser detection and ranging), they operate on the same basic principles as microwave radars. Because they operate at much shorter wavelengths, laser radars are capable of higher accuracy and more precise resolution than microwave radars. On the other hand, laser systems are subject to the vagaries of the atmosphere and

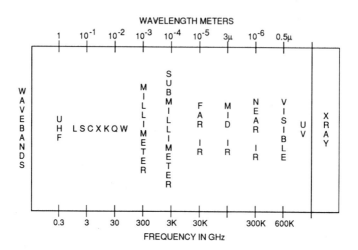

Figure 1.1 Frequency spectrum—UHF through x-ray bands.

are thus generally restricted to shorter ranges in the lower atmosphere than microwave radar. Rather than supplanting microwave radar, laser radar opens up new capabilities that exploit the great shift in wavelength, including—

- Tactical range and velocity imaging systems
- Autonomous missile guidance
- Precise aircraft navigation and guidance
- Precision fire control
- Remote atmospheric sensing

Modern laser radar systems combine the capabilities of radar and optical systems to allow simultaneous measurement of range, reflectivity, velocity, temperature, azimuth, and elevation angle. These six dimensions of target information can be utilized in fire control and weapon system applications to allow target acquisition, tracking, classification, and imaging. The modulation capabilities of microwave radar systems can be applied to laser transmitters to allow accurate target measurement and time/frequency gating of atmospheric or terrain background clutter. The optical resolution associated with laser systems results in a very small angular beamwidth to allow imaging, aimpoint assessment, precise target tracking, and autonomous operation.

Microwave radar systems typically have angular beamwidths that degrade tracking of ground targets in multipath or high-clutter conditions. The need for low-altitude flight results in a requirement for detection and avoidance of obstacles and power transmission lines, along with the ability to perform terrain following. In these low-altitude conditions, the detectability of microwave transmissions, low microwave off-axis wire cross sections, coupled with atmospheric and terrain clutter cross section, results in the consideration of laser radar systems.

Similarly, combining laser radar systems with passive sensors can result in a direct target size determination through range measurement or clutter rejection improvement through the use of laser radar signature and scan field capabilities. Where passive IR systems are dependent upon target emissivity and temperature characteristics, the laser radar is target reflectivity dependent. The combination of these physical attributes allows for reliable target measurement independent of time of day or target operation. Range and velocity measurements of stationary and moving targets allows the target to be segmented from the background to yield major improvements in automatic target detection and recognition systems.

1.2 RANGE EQUATION

Because the subject is electromagnetic propagation, the microwave radar range equation still applies:

$$P_R = \frac{P_T G_T}{4\pi R^2} \times \frac{\sigma}{4\pi R^2} \times \frac{\pi D^2}{4} \times \eta_{ATM} \eta_{SYS} \tag{1.1}$$

where

- P_R = received signal power (watts), or P_{SIG} later
- P_T = transmitter power (watts)
- G_T = transmitter antenna gain $\frac{4\pi}{\theta_T^2}$
- θ_T = transmitter beamwidth = $K_a \lambda / D$
- σ = effective target cross section (square meters)
- K_a = aperture illumination constant
- R = system range to target (meters)
- λ = wavelength (meters)
- D = aperture diameter (meters)
- η_{ATM} = atmospheric transmission factor
- η_{SYS} = system transmission factor

Substituting the above in Equation (1.1) results in

$$P_R = \frac{P_T \sigma D^4 \eta_{ATM} \eta_{SYS}}{16 R^4 \lambda^2 K_a^2} \text{ in the far field} \tag{1.2}$$

This equation shows that the received power is a function of transmitter power, the directional gain of the aperture, the effective scattering target cross section (σ), the target range, wavelength of operation, antenna/aperture size and the atmospheric and system transmission factors. The transmitter aperture gain may be expressed by the steradian solid angle of the transmitter beamwidth $(\theta_T)^2$ to that of the solid angle of a sphere which is equal to the relation $4\pi/\theta_T^2$. For laser beamwidths on the order of 1 mrad, the typical aperture gain at laser wavelengths is about 70 dB. The standard range equation applies only in the *far field* of the aperture. At typical microwave bands of $\lambda = 1$ to 10^{-3} m, the far-field distances are quite short, as shown in Figure 1.2. The far-field (*Fraunhofer*) region of an aperture is typically concerned with the distance $2D^2/\lambda$ [1] to infinity; in this vicinity, the generalized range equation applies. In some cases, the far-field distance occurs within the feed horn assembly of a microwave antenna. As illustrated by the figures, at a 10-μm wavelength, a 1m aperture has a far-field distance of approximately 200 km. As a result, it is not unusual to operate in the *near field* of the optical system; thus, modification of the range equation to account for near-field operation is required. This near-field effect modifies the beamwidth such that

$$\theta_T = \left[\left(\frac{K_a D}{R}\right)^2 + \left(\frac{K_a \lambda}{D}\right)^2\right]^{1/2} \tag{1.3}$$

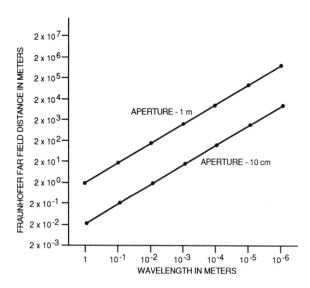

Figure 1.2 Far field distance *versus* wavelength for 1m and 10-cm apertures.

The effective target cross section is defined as

$$\sigma = \frac{4\pi}{\Omega} \rho_T dA \qquad (1.4)$$

where

Ω = scattering steradian solid angle of target
ρ_T = target reflectivity
dA = target area

Figure 1.3 shows that corner reflectors at microwave wavebands are usually very large; a cross section of 10^6 m² would require a 2.2m corner reflector at a 1.0-cm wavelength.

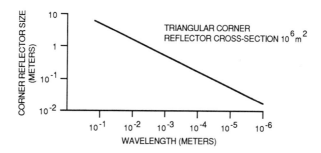

Figure 1.3 Triangular corner reflector size *versus* wavelength for 10^6 m² cross section.

Obtaining this same cross section at a 1-μm wavelength requires a corner reflector having a size of 2.2 cm (basically the size of a nickel). This effect is often used at optical wavebands to enhance cooperative target cross sections; very small corner reflector sizes can yield very high target gains. This may be seen in Equation (1.5) where

$$\sigma_{C.R.} = \frac{4\pi D^4}{3\lambda^2} \qquad (1.5)$$

If a mirror type target is used instead of a diffuse target, a transmitting beamwidth of 1 mrad ($\Omega = 10^{-6}$) would yield an approximate 70-dB increase in target cross section. This gain would occur when a mirror target was aligned properly to the receiver to allow the specular reflection from the target to be redirected toward the receiver. Because typical targets have a probabilistic distribution associated with their specular facets, the likelihood that the signal would return back to the receiver is very low. As a result, wide signal fluctuation is seen to exist from such a target as the mirror surface changes its angular orientation.

Diffuse targets, on the other hand, have large scattering fields. Physicists or those in the optical sciences tend to replace Ω with the value associated with the standard scattering diffuse target (*Lambertian target*) having a solid angle of π steradians, thereby reducing Equation (1.4) to

$$\sigma = 4\rho_T dA \qquad (1.6)$$

This situation may be compared to the standard isotropic target, used by the microwave system engineers, having a 4π scattering steradian angle ($\Omega = 4\pi$). For a diffuse extended radar target, scattering may only occur into a half sphere, Ω would then be equal to 2π, and σ becomes

$$\sigma = 2\rho_T dA (1/2 \text{ plane}) \qquad (1.7)$$

1.2.1 Range Equation Dependence on Target Area

The area (dA) illuminated by a circular aperture at a target range

$$dA = \frac{\pi R^2 \theta_T^2}{4} \text{ (target normal to beam)} \qquad (1.8)$$

If the target intercepts the entire beam, it is classified as an *extended target* which has a range square dependency, shown in Figure 1.4 and Equation (1.10), and all of the incident radiation is involved in the reflection process. However, not all targets meet this condition. A *point target* is one where the target area is smaller than the transmitter footprint; hence there is no range dependency in the scattering area. A *linear target* such as a wire can have a length larger than the illuminated area but a smaller width (d). This target then has an area having a linear dependency on range (Equations (1.11) and (1.12)).

Thus, for an extended Lambertian target,

$$\sigma_{EXT} = \pi \rho R^2 \theta_T^2 \qquad (1.9)$$

resulting in

$$P_R = \frac{\pi P_T \rho_T D^2}{(4R)^2} \eta_{ATM} \eta_{SYS} \qquad (1.10)$$

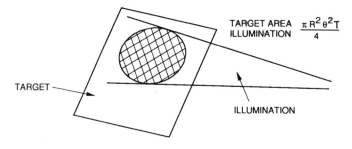
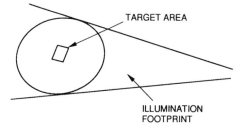
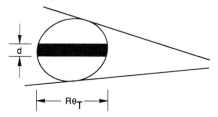

Figure 1.4 Range dependency of typical targets.

With narrow laser beams, the standard inverse fourth power of range targets for microwave systems may become an inverse R^2 extended target at the optical wave band. Other received power inverse range-dependent functions ($1/R^4$, $1/R^3$, et cetera), developed similarly in the microwave region, can also occur in the optical region. An inverse R^3 relationship may be obtained by considering a diffuse (Lambertian) wire target having wire diameters (d) and length $R\theta_T$. The target cross section may be shown to be approximately

$$\sigma_{wire} = 4\rho_{wire} R\theta d \tag{1.11}$$

and the range equation becomes

$$P_R = \frac{P_T \rho_{wire} dD^3 \eta_{SYS} \eta_{ATM}}{4R^3 K_a \lambda} \quad (1.12)$$

If the radar measurement involves collection of all the energy reflected from the illuminated spot, as in target detection, then the entire illuminated area is used in the target cross section calculation. Similarly, if an elemental area is to be used, such as in imaging, then the imaging cross section is used.

For a Lambertian diffuse point target, the cross section becomes

$$\sigma_{PT} = 4\rho_T dA \quad (1.13)$$

Substituting the point target cross section in the range equation results in the received signal power

$$P_R = \frac{P_T \rho dA dD^4}{4R^4 K_a^2 \lambda^2} \eta_{SYS} \eta_{ATM} \text{ (point target)} \quad (1.14)$$

Equation (1.2) assumes that the transmitter and receiver optics have the same diameter. Equation (1.15) expresses the typical optical range equation in terms of far-field radiance where the range from transmitter and receiver may not be equal.

Here the range equation expresses the received energy (power) as a function of the transmitter energy (power), output beam divergence, range from transmitter, target reflection characteristics, range from receiver, optics/atmospheric transmission, and receiver collection aperture. A basic form is shown in Equation (1.15) and Figure 1.5.

$$E_R = \frac{E_T}{R^2 \Omega_T} \frac{(\rho A_R)}{\Omega_R} \frac{A_C}{R_R^2} \eta_T \eta_R \quad (1.15)$$

where

$P_R(E_R)$ = received return power (energy) from target
$P_T(E_T)$ = transmitted laser power (energy)
R = range from the transmitter to the target
R_R = range from the receiver to the target
Ω_T = solid angle of the transmitted beam
Ω_R = solid angle of the reflected beam
A_R = cross-sectional area of the reflecting target $\left(\frac{\pi}{4} R^2 \Omega_T^2\right)$

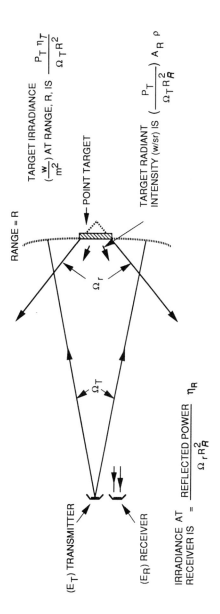

Figure 1.5 Laser range equation (radar range equation).

ρ = average reflectivity of the target (see Appendix A)
η_T = transmission coefficient for the transmitted beam (includes optics as well as atmospherics)
η_R = transmission coefficient for the reflected beam (optics and atmospherics)

1.3 DETECTION CONSIDERATIONS

In basic physics books, the dual nature of electromagnetic propagation is discussed, in the sense that electromagnetic fields may be seen to travel sometimes in a wave train and sometimes as a discrete packet of energy. The packets of energy are detected at laser wavelengths, whereas in the microwave region, the context is wave trains. The energy associated with one photon (this basic particle of light) is equal to hf, where h is Planck's constant (6.626×10^{-34} J-s) and f is the frequency.

It may be difficult to envision, but transmitter frequency can be related to energy. The number of photons (N_R) arriving in an optical receiver in a second (N_R/s) is the received power (P_R) divided by hf. In microwave terminology, the received signal power is in terms of watts which, in optics, is the number of photons arriving per second at a specific wavelength.

Figure 1.6 illustrates the energy associated with the photon as a function of wavelength. As the wavelength becomes shorter, the energy of a photon increases. Indicated with a dotted line is the standard thermal noise effect. Practically, it is impossible to detect a photon at microwave wavelengths—it is the minimum discernible signal in the electromagnetic spectrum—but thermal noise obscures the individual photon in that region. Thermal noise (kT) is derived from a term in physics called *blackbody radiation*. The blackbody radiation law [2] relates the power emitted from a blackbody radiator to the fourth power of the temperature of the radiator. The total blackbody radiation, P_{BBT}, may be expressed as the product of the radiant emittance, I_e, and the surface area within the field of view, A_S. Radiant emittance of a blackbody overall wavelength is expressed by the Stefan-Boltzmann law as

$$I_e = \sigma_T T^4 \qquad (1.16)$$

therefore

$$P_{BBT} = I_e A_S = \sigma_T T^4 A_S \qquad (1.17)$$

where

σ_T = Stefan-Boltzmann constant = 5.67×10^{-12} W cm^{-2} °K^{-4}
T = absolute temperature (Kelvin)°

Using a selective optical filter allows the receiving system to pass only a portion of P_{BBT}, thereby reducing the blackbody background noise contribution.

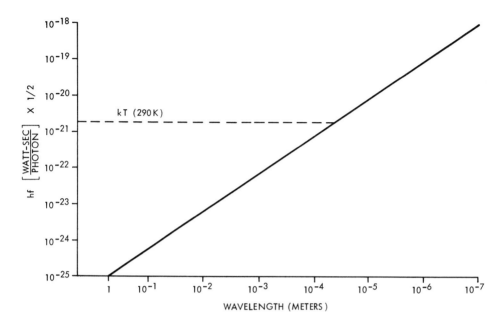

Figure 1.6 Photon energy *versus* wavelength.

The spectral radiant emittance of a blackbody into a hemisphere in the wavelength from λ to $\lambda + d\lambda$ may be expressed as

$$I(\lambda) = \frac{2\pi c^2 h}{\lambda^5} \frac{1}{\exp\left(\dfrac{hc}{\lambda kT}\right) - 1} d\lambda \tag{1.18}$$

where

$$2\pi c^2 h = 3.74 \times 10^{+4} \text{ W } \mu\text{m}^4 \text{ cm}^{-2} \tag{1.19}$$

$$\frac{ch}{k} = 1.438 \times 10^4 \ \mu\text{m } °\text{K} \tag{1.20}$$

$I(\lambda)d\lambda$ is the flux radiation into a hemisphere per unit area of the source, measured in watts per square centimeter.

$$I(\lambda)d\lambda = \frac{3.74 \times 10^4}{\lambda^5 \exp\left(\dfrac{1.438 \times 10^4}{\lambda T}\right) - 1} d\lambda \tag{1.21}$$

1.3.1 Noise Power Spectral Density

Quantum mechanical analyses [3] have shown that an ideal amplifier has a noise power spectral density referred to the input (Ψ) (watts/hertz) as follows:

$$\Psi(f) = \frac{hf}{\exp\left(\dfrac{hf}{kT}\right) - 1} + hf \tag{1.22}$$

where

T = absolute temperature of source
h = Planck's constant
f = transmitter frequency
k = Boltzmann's constant

Expanding $\exp(hf/kT)$ in the denominator of this expression in terms of an exponential series results in $1 + (hf/kT) + (1/2!)(hf/kT)^2 + \cdots$. When operating in the microwave band, kT is very large compared to hf, and the exponential series tends to $(1 + hf/kT)$. As a result, the hf terms in the denominator and numerator cancel, leaving kT—thermal noise. Thermal noise then is derived from Planck's blackbody radiation equation.

Plotting this relationship as a function of wavelength indicates that, at wavelengths longer than 10^{-4} m, thermal noise is the minimum discernible noise source; in the region ranging from that point toward shorter wavelengths, the minimum detectable signal is actually the photon noise.

The noise power spectral density *versus* wavelength is shown in Figure 1.7.

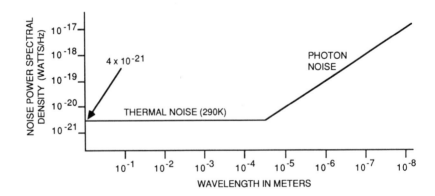

Figure 1.7 Noise power spectral density *versus* wavelength.

1.3.2 Receiver Detection Techniques

In Figure 1.8, diagrams are shown for incoherent and coherent detection receivers. The incoherent detection receiver at optical wavelengths is similar to a video radiometer receiver (i.e., an envelope detector at microwave wavelengths). However, an optical receiver has an additional term besides the signal term, P_{SIG}, the optical background power, P_{BK}, which is due to undesired signals such as sunlight, cloud reflections, and flares. The received signal competes with these external noise sources at the receiver. The received optical power, after suitable filtering, is applied to the optical detector; square law detection then occurs, producing a video bandwidth electrical signal.

The coherent detection receiver is similar to the incoherent; however, a portion of the laser signal f_o, is coupled to the optical detector via beamsplitters. As a result, the optical detector has the local oscillator power (P_{LO}) in addition to the received signal power, P_{SIG}, and the competing background terms, P_{BK}.

1.3.3 Background Noise Terms

Noise terms in an optical receiver are not the typical ones considered in the microwave receiver. As a result, major differences exist between optical receivers and microwave receivers. Background noise in optical receivers includes reflections of signals from the earth, the sun, the atmosphere, clouds, or any other source that contributes an undesired signal to the receiver. An analogy to this background effect is driving a car on a dark, foggy evening with the headlights on—a lot of light energy is scattered from the fog particles, producing a significant amount of backscattered fog energy as well as backscattered road light. As a result, the visibility of the road is limited. In this case, the backscatter from the fog represents an undesired signal because it masks the radiation returned from the ground, the desired signal. This same problem can occur for ungated laser systems propagating radiation into clear air, haze, and clouds.

Signal-induced noise refers to the shot noise caused by the received signal itself coming into a detector. The received signal is not only a source of desired radiation, it also causes a noise to be generated. This noise is called *quantum noise* (Poisson) because it is induced by the signal when the signal exists.

The following equations are those associated with calculating the amount of background radiation that may be incident upon a receiver.

Blackbody Radiation (P_{BB})

$$P_{BB} = \frac{\varepsilon \sigma_T T^4 \Delta \lambda \Omega_R A_R}{\pi} \eta_{SYS} \eta_{ATM} \qquad (1.23)$$

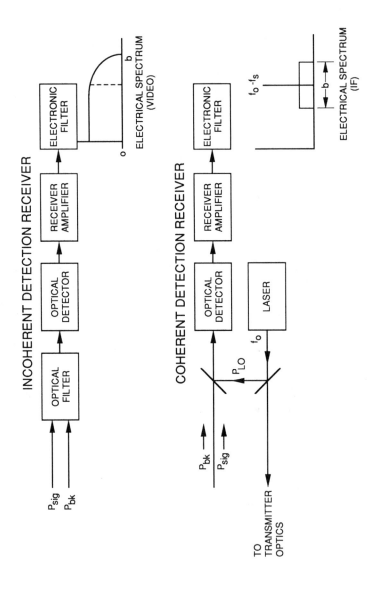

Figure 1.8 Receiver systems.

Solar Backscatter (P_{SB})

$$P_{SB} = k_1 S_{IRR} \times \Delta\lambda \Omega_R \rho \eta_{SYS} A_R \qquad (1.24)$$

Atmospheric Solar Scatter (P_{NS})

$$P_{NS} = k_1 S_{IRR} \times \Delta\lambda \Omega_R I_S \eta_{SYS} A_R \qquad (1.25)$$

where

- ε = target emissivity
- ρ = target reflectivity
- T = temperature (°K)
- $\Delta\lambda$ = optical bandwidth (micrometers)
- A_R = receiver area (square meters)
- k_1 = fraction of solar radiation penetrating Earth's atmosphere
- S_{IRR} = solar irradiance (W m^2-μm)
- I_S = atmospheric scatter coefficient
- η_{SYS} = system optical efficiency
- Ω_R = solid angle over which energy radiates from radiating body
- σ_T = Stefan-Boltzmann constant = 5.67×10^{-12} W cm^{-2} °K^{-4}

The blackbody radiation equation is a product of the radiant emittance, I_e, and the surface area within the field of view, A_S, has been discussed earlier under "Detection Considerations." The radiant emittance, I_e, is described by the Stefan-Boltzmann law as $\sigma_T T^4$; it includes the blackbody radiation over all wavelengths. The background noise impinging on a detector may be filtered optically to reduce the background energy competing with desired signal radiation. Figure 1.9 illustrates the

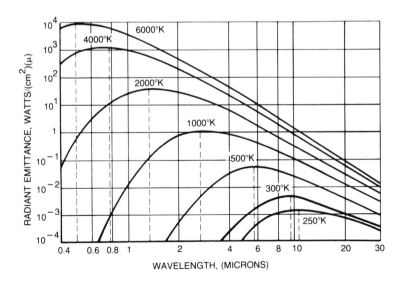

Figure 1.9 Blackbody radiant emittance *versus* wavelength.

radiant emittance as a function of wavelength for a variety of blackbody temperatures. As the temperature increases, the peak of the radiant emittance curve also increases. Figure 1.10 [4] shows the spectral irradiance of the sun (6000 °K) as a function of wavelength; these values can be used in the equations previously noted. Inasmuch as the energy from the sun may be either reflected or emitted from terrains, clouds, or targets, the reflectivities and emissivities of various terrain materials as a function of wavelength must be included in any noise calculation.

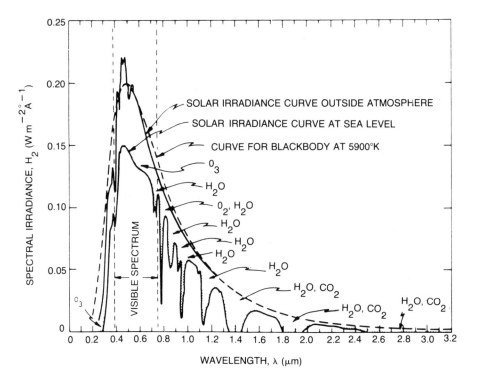

Figure 1.10 Spectral irradiance of the sun at mean earth-sun separation. (From *RCA Electro-Optics Handbook* [4].)

1.3.4 SNR Expression Development

Having discussed the receiver detection techniques and backgrounds, let us now turn to the derivation of the signal-to-noise ratio (SNR) expressions.

$$\text{SNR} = \frac{i_{SIG}^2}{i_{SN}^2 + i_{TH}^2 + i_{BK}^2 + i_{DK}^2 + i_{LO}^2} \quad (1.26)$$

where

i^2_{SIG}	=	mean square signal current
i^2_{SN}	=	mean square shot noise current
i^2_{TH}	=	mean square thermal noise current
i^2_{BK}	=	mean square background noise current
i^2_{DK}	=	mean square dark noise current
i^2_{LO}	=	mean square local oscillator noise current

The signal-to-noise expression above is shown to be related to the mean squared signal current squared, divided by the summation of noise current terms squared. As indicated above, the summation of noise terms involves shot noise, thermal noise, background noise, dark current, and (with a coherent detection system) a local oscillator noise. The photons or energy collected from the background result in a fluctuation of the carrier or electron densities in the detector, thereby contributing shot noise.

In the absence of photons at the detector, there is a current flowing, termed the *detector dark current*. Even though the detector surface may be blocked from having any radiation applied to it, its internal physics causes this leakage current to exist. Thermal-receiver noise, $1/f$ noise, and generation recombination noise can occur in both microwave and optical receivers. The thermal mean square noise current, which is conventionally referred to as the *receiver Johnson noise*, is expressed in terms of $4kTBNF/R_L$, where B is the electronic bandwidth, NF is the noise factor of the receiver following the detector and R_L is the detector load resistance. If a local oscillator signal is used for a coherent detection system, the local oscillator signal itself generates a shot noise similar to the received signal and background radiation. Expressions for these noise current terms are now given.

1.3.5 Receiver Noise and Signal Terms

Photons or energy collected from backgrounds result in a fluctuation of carrier or electron densities in the detector, and thereby contribute shot noise. The mean squared background noise term may be expressed as

$$i^2_{BK} = 2qP_{BK}\rho_i B \tag{1.27}$$

where

q	=	electron charge—1.602×10^{-19} coulombs
P_{BK}	=	background power (watts)
ρ_i	=	current responsivity (amps/watt)
B	=	electronic bandwidth (hertz)

Similarly, there is a fluctuation in the detector output caused by the random arrival of signal photons.

$$i_{SN}^2 = 2qP_{SIG}\rho_i BG^2 \tag{1.28}$$

Detector Dark Current (I_{DK})

$$i_{DK}^2 = 2qI_{DK}B \tag{1.29}$$

Thermal Noise Current

$$i_{TH}^2 = \frac{4kTBNF}{R_L} \tag{1.30}$$

where

NF = receiver noise factor
R_L = detector load resistance

Local Oscillator Induced Noise (Assuming Coherent Detection, Photovoltaic Detector)

$$i_{LO}^2 = 2qP_{LO}\rho_i B \tag{1.31}$$

where

P_{LO} = local oscillator power

For a photoconductor detector, the following noise term can arise:

Generation-Recombination Noise (Photo Conductors)

$$i_{GR}^2 = 4q\rho_i(P_{LO} + P_{SIG})B \tag{1.32}$$

The signal current is determined as

$$i_{SIG} = \frac{\eta_D q P_{SIG} G}{hf} \text{ incoherent} \tag{1.33}$$

$$i_{SIG} = \frac{\eta_D q \sqrt{2P_{SIG}P_{LO}}}{hf} \text{ coherent} \tag{1.34}$$

where

    η_D = detector quantum efficiency
    G = detector gain

The detector responsivity, $\eta q/hf$, is concerned with the conversion of optical power to receiver current and, as such, is represented by a term ρ_i, the current responsivity in amperes per watt.

Substituting these current expressions into the generalized SNR equation results in the incoherent and coherent system equations.

1.3.6 Comparison of Incoherent and Coherent Detection Receivers

The SNR equation for incoherent detection may be expressed as:

$$\text{SNR} = \frac{\eta_D P_{SIG}^2}{hf[2B(P_{SIG} + P_{BK})] + K_1 P_{DK} + K_2 P_{TH}} \qquad (1.35)$$

The SNR equation for coherent detection may be expressed as

$$\text{SNR} = \frac{\eta_D P_{LO} P_{SIG}}{hfB[(P_{LO} + P_{SIG} + P_{BK})] + K_3 P_{DK} + K_4 P_{TH}} \qquad (1.36)$$

where

    SNR = electrical signal power/electrical noise power
    η_D = detector quantum efficiency
    h = Planck's constant (6.626×10^{-34} J-S)
    f = transmission frequency
    B = electronic bandwidth
    P_{SIG} = received signal power
    P_{BK} = background power
    P_{DK} = equivalent dark current power = $\dfrac{A_d B}{(D^*)^2}$
    P_{TH} = equivalent receiver thermal noise $\dfrac{4kTBNF}{R}$
    P_{LO} = reference local oscillator power
    $K_1 = \dfrac{\eta_d}{\rho_i^2}$
    $K_2 = \dfrac{\eta_d}{\rho_i^2}$

k = Boltzmann's constant $\left(1.39 \times 10^{-23} \dfrac{\text{J}}{\text{°K}}\right)$

T = receiver temperature (290 °K)

$K_3 = K_4 = \dfrac{hf}{2qp_i}$

NF = receiver noise figure

R_L = resistance

where

A = detector area (square centimeters)

p_i = the detector current responsivity

q = the electron charge (1.6×10^{-19} coulombs)

D^* = specific detectivity (cm-Hz$^{1/2}$/W)

The SNR for the incoherent system has the received signal power squared in its numerator, and has a summation of noise terms associated with the return signal, the background signal, the dark current, and the thermal noise of the receiver in the denominator. The returned signal power and the background power are included as noise sources in the detection process because of the random photon arrival rate (Poisson noise). In the coherent detection system, the local oscillator power is an additional source of noise (compared to the incoherent system), and the numerator is related to the product of the received signal power and the local oscillator power. The local oscillator power is very important in the detection process; here, it may be increased so that it overwhelms all of the other noise sources. As a result, the local oscillator power in the denominator cancels out the local oscillator power in the numerator; the SNR is directly proportional to the received signal power, rather than to the received signal power squared (as with the incoherent system). Additionally, because the local oscillator power becomes the predominant noise source, the coherent detection system typically is background immune, since only signals that are phase coherent with the local oscillator are efficiently detected.

For coherent detection where the local oscillator power is increased to provide shot-noise-limited operation of the receiver, the SNR expression for coherent detection can be reduced to

$$\text{SNR} = \dfrac{i_{SIG}^2}{i_N^2} = \dfrac{\eta_D P_{SIG}}{hfB} \quad \text{or} \quad \dfrac{\eta_D E_{SIG}}{hf} \qquad (1.37)$$

where E_{SIG} is the received signal energy, B is the matched filter bandwidth $\left(B = \dfrac{1}{T}\right)$ and SNR represents the number of detected photons if $\eta_D = 1$.

For a background noise-limited incoherent receiver, Equation (1.35) becomes

$$\text{SNR} = \frac{\eta_D P_{SIG}^2}{2hfBP_{BK}} \qquad (1.38)$$

Figure 1.11 illustrates the reference transmitter power *versus* SNR relationship for coherent and incoherent detection laser radar systems using a typical 100-ns pulse width and the following parameters at 10.6 microns,

η_D = 0.5
hf = 1.9 × 10^{-20} joules

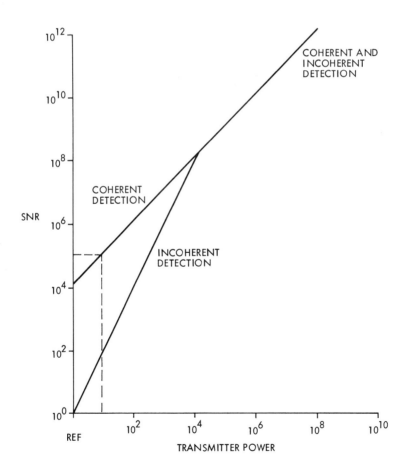

Figure 1.11 Transmitter power *versus* SNR.

$$D^* = 2 \times 10^{10} \frac{\text{cm } \sqrt{\text{Hz}}}{\text{W}}$$

$\sqrt{A_d} = 0.03$ cm

$\rho_i = 4$ A/W

$R = 1000$ ohms

It may be observed that as the SNR requirement increases, the transmitter power of the coherent system increases linearly, and that of the incoherent system increases as the square root. In the limit, incoherent detection systems approach the sensitivity of coherent systems for very large SNRs. For a typical SNR requirement of 100 (20 dB), the coherent system is seen to have a 30-dB increased sensitivity over that of an incoherent system.

1.3.7 Noise Figure

Having discussed the minimum discernible signal, attention naturally is turned to the *noise figure* of an optical receiver. As demonstrated thus far, the minimum discernible signals at microwaves and at optical regions are different and comparisons are difficult. However, if the noise figure of a coherent optical receiver were to be compared to the thermal noise in the microwave receiver, the following equations could be used.

The SNR power of a microwave receiver may be expressed as

$$\text{SNR}_M = \frac{P_{SIG}}{kT_s B} \tag{1.39}$$

where

P_{SIG} = received signal power
k = Boltzmann's constant (1.39×10^{-23} J/K)
T_s = system noise temperature
B = receiver bandwidth

The SNR power for a quantum noise-limited coherent optical receiver may be expressed as

$$\text{SNR}_O = \frac{\eta_D P_{SIG}}{hfB} \quad \text{or} \quad \frac{\eta_D E_{SIG}}{hf} \tag{1.40}$$

where

h = Planck's constant
f = transmission frequency

η_D = quantum efficiency
E_{SIG} = received signal energy

The effective noise figure of the coherent optical receiver at 10.6 μm compared to that of a 290 °K microwave receiver is

$$(NF_{EFF})_O = \frac{hf}{\eta_D kT_s} \tag{1.41}$$

at 10.6 μm, $hf = 2 \times 10^{-20}$, $kT_s = 4 \times 10^{-21}$, and $\eta = 0.5$. Therefore

$$(NF_{EFF})_O = 10 \tag{1.42}$$

These equations compare the SNR of the microwave receiver in terms of its minimum discernible signal (kT_s) to that of a quantum noise-limited optical receiver having a minimum discernible signal of hf. If a 50% quantum efficiency detector were used, the noise figure of the optical system compared to a microwave system having a 290 °K system temperature would be about 10 dB poorer at 10.6 μm; at shorter wavelengths, the noise figure would be even higher.

The term "quantum efficiency" relates the efficiency by which photons of light are converted to electrical signals by the detector. If one photon were to arrive on a detector surface and one photoelectron were to be emitted, the device would have 100% quantum efficiency. For the analysis used (that is, a quantum efficiency of 50%), on average two photons must arrive in order to generate one photoelectron in the receiving system. In effect, this receiver would have an optical noise figure of 3 dB relative to the minimum discernible optical noise.

1.4 BEAMWIDTH

The diffraction-limited beamwidth of a transmitting source (θ_T) may be expressed as

$$\theta_T = K_a \lambda / D \tag{1.43}$$

where K_a is a constant determined by the aperture illumination function and, if λ and D are in the same units, θ is in radians.

Obviously, for a fixed aperture size, as the wavelength decreases, the beamwidth decreases. For a 1-cm waveband and a 30-cm aperture, the beamwidth is approximately 33 mrad. Similarly, if the wavelength is decreased to 10^{-3} cm (10 μm), the beamwidth becomes 33 μrad for the same aperture size. A constant, K_a, is used in these expressions to illustrate the difference between radar and various optical beamwidth definitions. Optical beamwidths are often specified at the $1/e^2$ points, whereas the standard radar notation of beamwidth is typically measured at the half-power points. It is necessary, therefore, to be aware of the differences in beamwidth definition at different points in the electromagnetic spectrum. Figure 1.12 illustrates the aperture illumination effect on system beamwidth, and shows the result of illuminating an aperture with a Gaussian beam where the diameter is matched to the $1/e^2$, $1/e$, or $1/2$ power points of the Gaussian beam.

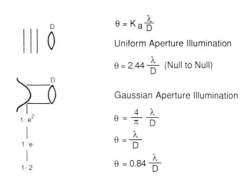

Figure 1.12 Aperture illumination function *versus* beam diameter (θ_T) [5].

If all-weather long-range radar system operation is desired, the system designer is forced to select long wavelengths. However, these systems must deal with the limitations due to multipath and angular resolution noted below:

- *Multipath*—Results in inaccurate tracking (typically in elevation)
- *Angular resolution*—Results in limited classification, separation, identification and aimpoint selection of targets

Because of these and other reasons, systems are developed which, although they are not all-weather, provide a significant adverse-weather capability. Currently, millimeter and infrared systems are being considered for this role. Operation in the millimeter waveband will allow target detection under more adverse weather conditions of advection (high moisture) fogs and battlefield obscurants than infrared systems. However, multipath [6–9] problems such as those that occur over snow-covered terrain, runways, water surfaces, and smooth terrain may prevent accurate target elevation tracking for ground-to-ground and low-altitude air-to-ground conditions, where accurate target tracking is required, while rain backscatter can provide an atomspheric clutter problem. Because snow-covered terrain can, statistically, occur for periods of time longer than poor visibility conditions, and aimpoint selection is required for some systems, optical portions of the electromagnetic spectrum are considered for these radar-type applications.

The basic multipath effect is caused by a transmission of microwave energy to a target shown in Figure 1.13, at a low elevation angle. The microwave radar beamwidth, θ_B, is sufficiently wide that some of the energy from the antenna system illuminates the terrain or water in addition to the target. As a result, some of the reflected energy from the ocean or background is reflected to the target and subsequently returned to the microwave radar system. In such a reflection pattern, the monopulse receiver sum signal increases and decreases, depending upon the phase of that reflected energy.

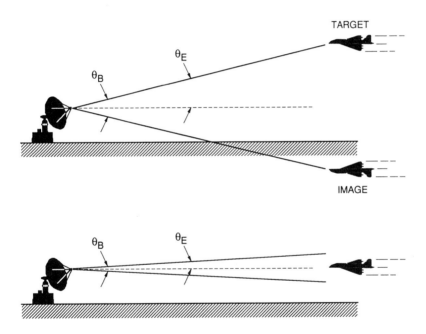

Figure 1.13 Multipath errors.

The energy reflected from the surface causes the difference channel of the tracking system to have an error signal caused by the signal return from the ocean or the terrain. To the degree that the signal returned from the lower path is within the view of the receiver main beam or side lobes of the microwave antenna, it provides an error source. This error source causes the microwave antenna system to have significant errors in the elevation track, depending on the surface backscatter effect.

Correspondingly, to the degree that the surface reflection factor is symmetrical about the microwave receiver, this effect is less prevalent in the azimuth channel. As a result, a laser system search field can be configured that is broader in elevation than it is in azimuth. The use of lasers and the corresponding wavelength associated with them results in a very much narrower beamwidth. Therefore, more of the electromagnetic energy propagated from the laser can be directed to the target, and the multipath errors are subsequently effectively eliminated.

1.5 SEARCH FIELD

The narrow beamwidths associated with laser radars necessitate very high data rate transmitters if large search fields are required. The solid angle to be searched is

$$\Omega_S = \frac{A_S}{R^2} \tag{1.44}$$

where

A_S = area to be searched
R = range to target

By dividing the solid angle to be searched (Ω_S) by that associated with a diffraction-limited transmitting aperture ($\Omega_{XMIT} = (\lambda K_a/D)^2$, the number of cells ($N$)$N = \Omega_s/\Omega_{XMIT}$ to be investigated may be determined. For a 1m circular aperture diameter, the transmitter solid angle field varies from 10^{-10} to 10^{-14} sr over a wavelength range of 10 to 0.1 μm, respectively. Correspondingly, the number of beam positions in the search field increases as the square of the wavelength decreases; that is, a YAG laser (1.06 μm) search field will have 100 more beam positions, for the same aperture diameter, in a given search field than that of a CO_2 laser (10.6 μm). Assuming a target location uncertainty of 1 by 1 mrad, the number of beam positions (N) to be interrogated with systems having a diffraction-limited aperture of 1m or 0.15m and $K_a = 1$ are

λ Wavelength (μm)	$N(D = 1m)$ No. of Beam Positions	$N(D = 0.15m)$ No. of Beam Positions
10	10^4	225
1	10^6	2×10^4
0.1	10^8	2×10^6

The time required to search a field (Frame Time (F.T.)) may be expressed as

$$F.T. = N(T.O.T.) \tag{1.45}$$

where

    N = number of cells
    $T.O.T.$ = time on target or measurement interval time

Here it may be readily observed that laser radars require high repetition rates or long acquisition times in order to perform the target search function, unless multiple beams are used. However, the angular resolution associated with a laser radar, coupled with the modulation bandwidth capability, allows substantial target measurement capability, such as simultaneous angle, range, velocity, and intensity during a single measurement.

1.6 SEARCH FIELD FIGURE OF MERIT

Typical figures of merit for microwave radar search systems involve the power-aperture area product. The larger this ratio the more capable is a radar system to scan a large field in a given time. The figure of merit for a coherent laser radar system is not identical to that of a microwave system and it is wavelength dependent.

The steradian search field (Ω_s) may be approximated by

$$\Omega_s = \frac{F.T. \cdot \theta_B^2}{T.O.T.} \tag{1.46}$$

where

    $F.T.$ = frame time
    θ_B = beamwidth
    $T.O.T.$ = time on target

Radar

$$\text{SNR} = \frac{P_R}{kTBNF} \tag{1.47}$$

Coherent Laser Radar

$$\text{SNR} = \frac{\eta_D P_R}{hfB} \tag{1.48}$$

where

P_R = received signal power
k = Boltzmann's constant
T = temperature
B = bandwidth
NF = receiver noise figure
η_D = detector quantum efficiency
h = Planck's constant
$f = \dfrac{c}{\lambda}$ = frequency

assuming the system bandwidth is matched to the time on target

$$B = \frac{1}{T.O.T.} \tag{1.49}$$

results in the search field for the detection of a point target becoming

Radar

$$\Omega_S = \frac{F.T.\, P_T\, \sigma\, D^2\, \eta_{SYS}\, \eta_{ATM}}{16\, \text{SNR}\, kT\, NF\, R^4} \tag{1.50}$$

Laser

$$\Omega_S = \frac{F.T.\, P_T\, \sigma\, D^2\, \lambda\, \eta_{SYS}\, \eta_{ATM}}{16\, \text{SNR}\, hc\, R^4} \tag{1.51}$$

letting the detector quantum efficiency be equivalent to the microwave receiver noise figure,

$$\eta_D = \frac{1}{NF} \tag{1.52}$$

If the frame time, target cross section, system efficiency, SNR requirement, and range are the same for both radar and laser systems, the search field equations may be parametrically expressed as

Radar

$$\Omega_S = C_1 P_T D^2 \qquad (1.53)$$

Laser

$$\Omega_S = C_2 P_T D^2 \lambda \qquad (1.54)$$

where C_1 and C_2 are constants for the assumptions made, and results in the conclusion that the laser radar system search field figure of merit $(P_T D^2 \lambda)$ is wavelength dependent; that is, for similar power-aperture area products (for coherent reception), the longer wavelength systems have greater scan field capability. In this example, the laser footprints were assumed to be contiguous. Increased search field capability may be obtained by using multiple transmitter and receiver beams, and/or beam ellipticity when targets are not normal to the beam, or by spacing the laser footprints in a noncontiguous manner, consistent with target size and detection requirements.

1.7 SYSTEM OPTIMIZATION

1.7.1 Incoherent Receiver

In order to optimize system SNR, for an incoherent receiver, the system engineer must choose wisely from a wide variety of variables. The equation below indicates some of the considerations involved with detector selection and illustrates that, if the optical background energy is the limiting noise source after suitable filtering, any choice should be predicted upon the device having high quantum efficiency (in which case the SNR is increased).

$$\text{SNR} = \frac{\eta_D P_{SIG}^2}{2hfB \, [P_{BK} + P_{SIG}] + K_1 P_{DK} + K_2 P_{TH}} \qquad (1.55)$$

Decreased by narrow bandpass filter, narrow field of view, lower background emission

NEP
Smaller detector— smaller dark current; higher responsivity— lower noise equivalent power (NEP)

Select best noise figure: cool amplifier large load resistance

where

    ρ_i = responsivity (current)
    I_{DK} = detector dark current (discussed in Chapter 4)

Correspondingly, if the detector dark current is the limiting noise term, the detector having the smallest NEP should be chosen. Detector NEP may be reduced by choosing small area or high responsivity detectors. Photomultiplier detectors have significant amplification that can reduce the effect of detector dark current. These will be discussed in Chapter 4. Similarly, the choice of a receiver noise figure or load resistance consistent with other receiver noise expressions is required to ensure proper system operation and reduce the thermal noise.

Equation (1.1) illustrated that the received signal power P_R (or P_{SIG}) was quadratically related to the round-trip atmospheric transmission, and it may be seen in Equation (1.55) that the incoherent SNR expression is related to the *signal power squared*. When the receiver is limited by the detector dark current or thermal noise, the incoherent system will be a function of the one-way atmospheric transmission to the *fourth power*. Similarly, under this condition the SNR can have range dependency of $1/R^8$, $1/R^6$, or $1/R^4$ for a point, linear, and extended target, respectively.

1.7.2 Coherent Receiver

For a coherent receiver, the SNR (for a point target) may be expressed as

$$\text{SNR} = \frac{E_T \sigma D^4 \eta_{SYS} \eta_{ATM}}{16 \lambda R^4 h C} \qquad (1.56)$$

Here it may be seen that the optics diameter is linearly related to range performance; that is, a doubling of optics diameter doubles the range performance with the exception of atmospheric transmission (all other items being equal).

1.7.3 System Efficiency

The system efficiency term in the signal-to-noise equation includes a variety of losses. These losses include those typically associated with radar systems in general (throughput loss, mismatch, bandwidth mismatch, *et cetera*), and some unique to optical systems. In this chapter, the discussions will include an itemization of system loss considerations, including the effects of temporal and spatial coherence on laser systems.

The following itemize some of the elements involved in the considerations of system efficiency:

    η_{OPTICS} Transmission and reception losses relating to the laser beam being coupled through or reflecting from elements in the optical train.

η_{TARGET} The reflectivity of a target surface determines the percentage of energy reflected (see Appendix A for typical material reflectivity for a variety of wavelengths), and the surface characteristics will determine the solid angle of the reradiation field. Surface roughness will determine the specularity or diffusivity of the target and, subsequently, the detection SNR requirements. The detection probability requirements for specular and diffuse targets are discussed in Chapter 3.

$\eta_{SPECKLE}$ Speckle effects [10,11].

When a coherent laser source illuminates a diffuse target, the scattered signal from the object is observed to have a granular spatial structure that has been described as "speckle." Speckle is caused by the coherent waves scattering from points on the target interfering with each other. When the waves interfere constructively or destructively, a bright or dark spot is correspondingly observed with the human eye for a visible source. Correspondingly, the speckle occurs at the received aperture and affects the detection process for both coherent and incoherent detection receivers. This target-induced interference phenomenology causes the standard deviation of the signal to be equal to the mean value and provides 100% modulation of the received signal. This modulation level may be reduced by averaging a number (N) of independent measurements, which will reduce the modulation by $\sqrt{N}$.

For a coaxial system sharing the same aperture, the speckle effect results in a diffuse target loss of 3 dB (0.5).

$\eta_{DETECTOR}$ Detector efficiency determines how well optical energy is converted to electrons, and typically it is stated as the quantum efficiency or related to the minimum discernible signal level described by the detector (NEP) noise equivalent power. (See Chapter 4 for values.)

$\eta_{POLARIZATION}$ Transceiver configurations have different sensitivity to the polarization characteristics of targets, and care must be exercised in the assumption of the target polarization retentiveness [12]. Table 1.1 [13] shows that these losses could vary between 1 dB for like transceiver polarization to 10 dB for crossed polarization, at a 10.6-μm wavelength.

$\eta_{ELECTRONIC}$ Receiver/processor loss. Having converted the optical signal to electrons, standard system losses need to be addressed. The gain and noise figure of the receiver amplifiers determine the loss due to Johnson noise. Protection of the receiver from the transmitter energy frequently requires gating of the receiver sensitivity on and off, or sensitivity time controlled programming of the receiver gain. Some of the typical losses include—

Table 1.1
Depolarization for Various Targets Relative to Horizontal Transmitter and Receiver Polarization (Courtesy of Applied Optics)

Polarization State				Target		
Transmitter	Receiver	Concrete Block Wall	Orange Trees	Dirt Bank	Power Pole (914 m/σ)	Eucalyptus Trees
Vertical	Horizontal	20%	(a)	18%	21%	8%
Vertical	Vertical	85%	80%	91%	120%	90%
Horizontal	Vertical	20%	(b)	24%	14%	6%

(a) Too small to record.
(b) Values greater than 100% result from the arbitrary choice of a horizontal polarization reference.

- Range gate loss due to gating of the receiver sensitivity, sampling only a portion of the received signal, or being open too long and resulting in more noise integration in the processor
- Bandwidth mismatch losses caused by the predetection and postdetection filters not matching the signal bandwidth
- Detection integration loss
- Straddling loss—loss associated with the received signal not being centered in the matched filter, but straddling multiple filters

$\eta_{WAVEFORM}$ Modulation of the laser (to provide waveforms allowing information extraction by a receiver/processor) can reduce the effective average power of the system by inducing nonlinearities, unwanted signals, bandwidth matching inefficiencies, and modulation loss.

$\eta_{BEAMSHAPE}$ Aperture beamshape loss. System calculations typically assume that the target is directly on axis and receives the maximum aperture gain. Dependent upon beamwidth definition half power, $1/e$, and $1/e^2$, these losses may be (1–2 dB) for point targets.

$\eta_{TRUNCATION}$ Truncation loss is related to aperture illumination function. Frequently, the theoretical analysis of the aperture illumination assumes an infinite aperture to simplify calculations. Realistic apertures result in truncating the laser beams and can typically result in truncation losses of several decibels.

$\eta_{ATMOSPHERE}$ Propagation through the atmosphere results in laser beam attenuation due to scattering and absorption (see Chapter 3). Refractive index changes caused by turbulence may limit the receiver aperture size, affect signal statistics, and determine integration time constant requirements in the signal processor.

$\eta_{COHERENCE}$ These losses relate to the temporal and spatial coherence requirement of the transceiver [10,14].

1.8 TEMPORAL AND SPATIAL COHERENCE

Incoherent sources such as blackbody radiation have wide bandwidth and, subsequently, have very small time and distance over which the beam may maintain a coherent temporal phase relationship with itself. Laser sources can have substantially narrow linewidths, allowing long-range coherent operation. With the advent of semiconductor pumping of solid-state lasers and the narrow linewidths inherent in gas lasers, coherent operation is feasible throughout the optical spectrum from a source point of view. Correspondingly, because these lasers have high frequency purity, the optical beams can obtain diffraction-limited performance. Utilization of the receiver processing bandwidth matched to the bandwidth of the source (instantaneous linewidth plus any change in linewidth over the propagation time) will result in the receiver achieving maximum coherent sensitivity. Similarly, the receiver's ability to collect coherent energy imposes spatial limitations on the transceiver configuration, atmospheric propagation, and detection interferometer alignment.

Temporal and spatial coherence of an electromagnetic wave is determined by the time or spatial interval over which an electromagnetic wave can be in phase with itself. This may be expressed (see Wolf and Zissis, p. 23 [15]) in terms of the bandwidth of the source or receiver.

Temporal Coherence

$$\Delta t = 1/\Delta F \qquad (1.57)$$

Δt = time over which the signal is coherent
ΔF = bandwidth of the source

Spatial Coherence

$$\Delta R_C = C\Delta t = C/\Delta F \qquad (1.58)$$

C = speed of light in a vacuum
ΔR_C = distance over which the signal is coherent

In order to use a coherent detection system, temporal and spatial coherence requirements must be maintained. Spatial photomixing requirements necessitate that the local oscillator and received signal optical phase front be aligned over the active detector surface area. Figure 1.14(A) uses a laser transmitter, emitting a spatial and temporal coherent signal coupled through an amplifier chain and transmitted to the atmosphere as $E(t) = B \cos\omega_{oT}$. This signal backscattered and Doppler shifted is then collected by the receiver optics in the form $E_s = C \cos(\omega_o \pm \omega_d)\tau$. After suitable receiver optical transmission, it impinges on the receiver detector area, along with a local oscillator signal, $E_{LO} = A \cos\omega_o\tau$, derived from the reference transmitting laser.

The signal power derived from the cross-product term is determined by standard range equation relationships as well as the degree to which temporal and spatial coherences and optical alignments have been maintained. The requirements for photomixing spatial coherent detection are illustrated in Figure 1.14(B) [16]. Phase front distortion due to atmospheric turbulence perturbation of the refraction index can result in limitations on the receiver aperture area of a coherent detection system.

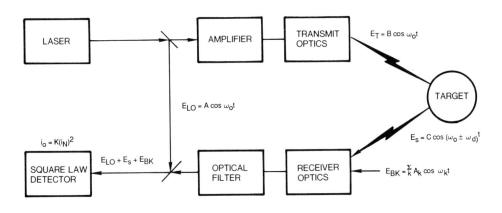

Figure 1.14(A) Coherent detection system.

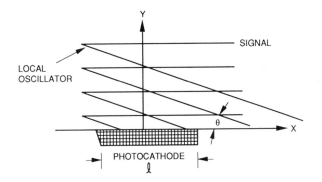

FROM MIXING PROCESS THE SIGNAL TERM MAY BE SHOWN TO BE:*

$$P_S P_{LO} \cos W_{if} \tau \left[\frac{\sin X}{X}\right]$$

WHERE $X = \dfrac{\pi \ell \sin\theta}{\lambda}$

1) IF $X \gg 1$ THEN THE SIGNAL TERM IS DECREASED IN AMPLITUDE

2) IF $\dfrac{\sin X}{X} = 1$ THEN THE SIGNAL TERM IS OPTIMUM

Figure 1.14(B) Photomixing spatial requirements for coherent optical detection system. (From Ross, M., *Laser Receivers*, Wiley & Sons. Used with permission.)

1.8.1 Spatial Coherence

The area over which the received signal is coherently detected was found by Forrestor [17] to be determined by the operating wavelength (λ) aperture area (A_a) and solid angle (Ω):

$$\Omega = \frac{\lambda^2}{A_a} \qquad (1.59)$$

and the best coherence efficiency occurs when the transmitter and receiver have matched fields.

This expression basically determines that the selection of the diffraction-limited transceiver aperture size for a system has an impact on the system efficiency and limits the instantaneous field.

1.8.2 Spatial Coherence Alignment

Optical heterodyning occurs when the received signal and local oscillator waves are superimposed on a detector surface. Beard and Fried [18] showed that, if the return signal was focused onto the detector and combined with a collimated local oscillator beam that critical alignment requirements between the local oscillator signal and the received signal spatial wavefronts oscillator might be reduced. In their analysis, the angular mismatch ($\Delta \theta$) that allowed for a $\lambda/4$ mismatch was

$$\Delta \theta \leq \pm \frac{D}{4F} \qquad (1.60)$$

where

D = diameter of the detector focusing lens
F = focal length of detector focusing lens

However, this technique did require an accurate placement of a local oscillator aperture. Subsequent system improvements resulted in both the local oscillator and received signal being combined through the focusing lens, with wavefront matching being performed on a beam recombining beamsplitter.

1.8.3 Coherent System Alignment Requirements [19,20,21,22]

The efficiency at which the electromagnetic coherent beams are heterodyned (mixed) on a detector has been expressed by Dr. K. Seeber [19] as the heterodyne efficiency

(η_{HET}) when

$$\eta_{HET} = \exp\left[-\left(\frac{\sigma_H}{\lambda/D}\right)^2\right] \qquad (1.61)$$

Selecting an allowable heterodyne loss η_{HET} of 3 dB (0.5) due to optical heterodyne beam misalignment, σ_H, results in $\sigma_H = 0.83\ \lambda/D$.

Figure 1.15 shows the heterodyne efficiency *versus* angular alignment requirements.

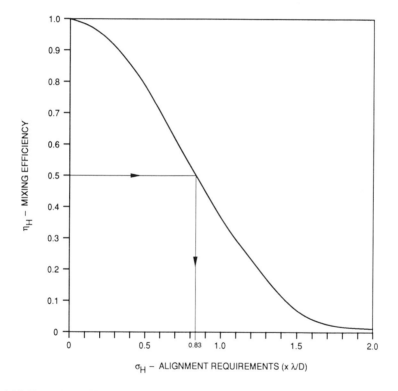

Figure 1.15 Heterodyne efficiency *versus* angular alignment requirements.

1.8.4 Angular Scan Rate—Lag Angle Consideration

The angular scan rate (θ) of systems may be limited by the time taken to propagate to the target and back being longer than the time for a transmitted beam to move one-half of a beamwidth. Studies conducted by Dr. K. Seeber [19] have shown that

$$\dot{\theta} = \frac{0.5\lambda C}{2DR} \qquad (1.62)$$

where

D = optics diameter
R = range
C = speed of light
λ = wavelength

This is further illustrated in Figure 1.16.

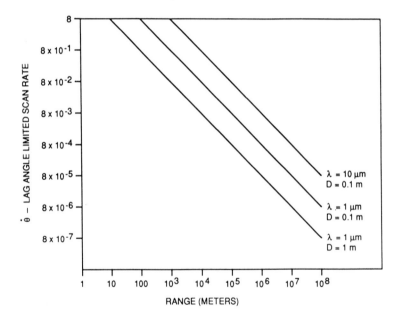

Figure 1.16 Lag angle scan limitations.

1.8.5 Effective Aperture Diameter for Coherent Receivers

Perturbations in the refractive index of a propagation medium caused by turbulence result in limitations on the receiver aperture area of a coherent detection system [23–27]. The effective aperture (D_{EFF}) represents the actual aperture diameter at which the heterodyne signal is reduced 3.0 dB below the level expected in the absence of turbulence:

$$D_{EFF} = \left(0.0588 \frac{\lambda^2}{C_N^2 R}\right)^{3/5} \qquad (1.63)$$

where

C_N^2 = atmospheric structure function for index of refraction variation (centimeters$^{-2/3}$)
λ = wavelength (centimeters)
R = path length (centimeters)
D_{EFF} = effective optics diameter (centimeters)

The structure function coefficient function, C_N, as a function of height may be observed in Figure 1.17.

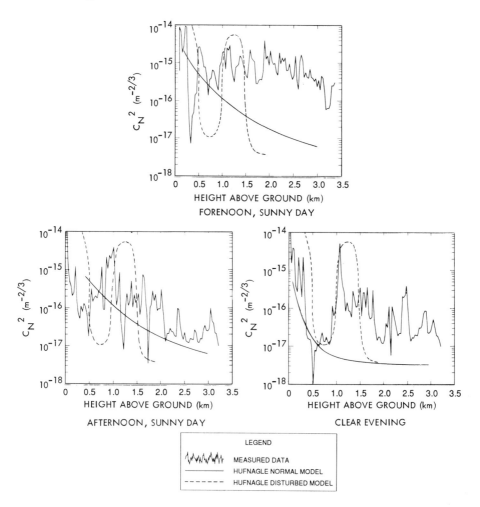

Figure 1.17 Structure function variations with altitude. (Courtesy of Lawrence et al. [23]; see also Brookner [24].)

1.8.6 Temporal Coherence

Incoherent radiation sources (blackbody, sun, *et cetera*) have high spectral content (large bandwidth—ΔF), which results in degraded phase coherence when attempts are made to correlate the emitted signal with itself (Equation 1.57).

Laser sources tend to have a high spectral purity, resulting in narrow line emissions and long coherence time. The coherence time for an electromagnetic source may be related to distance via knowledge of the speed of an electromagnetic wave (approximately 3×10^8 m/s), in radar applications where the round-trip propagation over a distance R to a target results in a coherence distance (Equation 1.58) of

$$\Delta R_C = \frac{C}{2\Delta F} \qquad (1.64)$$

The purity of the frequency of the transmitted signal when compared with itself at a later time (t) will have components of the initial spectral linewidth and the change in frequency over the propagation time. This combination of instantaneous linewidth and change in frequency over the propagation time determines the minimum coherent processing bandwidth of the receiver electronics. For example, suppose the instantaneous linewidth of a laser was 1 kHz (Figure 1.18 shows some experimental measurements as measured by beating the signal against itself in the laboratory [28,29]), and, when measured over a round-trip path of 10 km, the spectrum width combined with center frequency instabilities (uncorrected) resulted in a 10-kHz spectrum width, as measured in a coaxial transceiver.

$$\Delta t = \frac{1}{1 \text{ kHz}} = 10^{-3} \text{ sec in the laboratory and} \qquad (1.65)$$

$$\Delta t = \frac{1}{10 \text{ kHz}} = 10^{-4} \text{ sec over a 10-km path} \qquad (1.66)$$

In this case the laser has sufficient instantaneous linewidth to support round-trip range propagation of

$$\Delta R = \frac{C}{2\Delta F} = \frac{1.5 \times 15^8}{10^3} = 1.5 \times 10^5 \text{ m} \qquad (1.67)$$

but transmitter instabilities with time reduce our ability to measure this accurately beyond 15 km. Knowledge, measurement, and correction of the frequency instability could potentially be used to restore the measurement capability to 150 km.

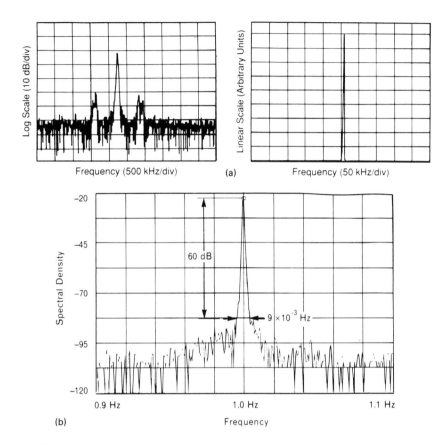

Figure 1.18 Laser linewidth measurements. (a) Nd: YAG microchip laser stability, (b) phase locked CW CO_2 laser stability. (Courtesy of MIT Lincoln Laboratory.)

1.8.7 Scan Spectral Broadening

In the process of scanning an electromagnetic wave over a scan field, a time-changing phase change occurs over the scan aperture.

Brandeiwie and Davis [13] showed that when a laser beam diameter (D) is scanned by a flat mirror over an angle θ at an angular scan rate W_s, the RMS broadening of the laser frequency will be

$$\Delta F_{SB} = \sqrt{2}\left(\frac{D}{\lambda}\right)W_s\theta \tag{1.68}$$

This scan broadening is caused by the mirror motion.

1.9 MEASUREMENT ERRORS

Precise determination of a target's location may be obtained by applying a modulation waveform to the transmitted signal and subsequently processing the received signal through a matched filter for the waveform (Fourier transform).

Determination of when a received signal crosses a threshold value or assessment of the centroid location of a signal spectrum in a uniform noise (white) receiver has been shown to depend on the SNR and/or the effective bandwidth (B_e), effective time duration (T_e), or angular beamwidth (θ_T) of the configured system.

Skolnik [30] showed that the one-sigma RMS errors in time (σ_{Te}) and range (σ_R) could be expressed as

$$\sigma_{Te} = \frac{1}{B_e(2E/N_o)^{1/2}} \tag{1.69}$$

$$\sigma_R = \frac{C}{2}\sigma_{Te} \tag{1.70}$$

and that the one-sigma RMS error in frequency (σ_f) (Doppler velocity) could be expressed as

$$\sigma_f = \frac{1}{T_e(2E/N_o)^{1/2}} \tag{1.71}$$

where E is the signal energy, N_o is the noise power per cycle, and, in a bandwidth (B), the noise power (N_1) is (N_oB), with the noise considered uniform Gaussian (white).

The effective bandwidth (B_e), and effective time duration (T_e) are determined by the matched filter characteristics associated with the modulation waveform. For a perfectly rectangular pulse of width τ, B_e is equal to $1/\tau$ for an IF receiver while T_e is equal to $\pi\tau/\sqrt{3}$, resulting in

$$\sigma_R = \frac{C\tau}{2(2E/N_o)^{1/2}} \text{ (time domain)} \tag{1.72}$$

$$\sigma_f = \frac{\sqrt{3}}{\pi\tau(2E/N_o)^{1/2}} \text{ frequency domain} \tag{1.73}$$

The *uncertainty principle of radar* (D. Gabor, *Theory of Communication Journal of IEEE*, Vol. 93, Part III, 1946) illustrates that the product of the effective time duration (T_e) and effective bandwidth (B_e) is

$$T_e B_e \geq \pi \tag{1.74}$$

This was utilized by J.F. Marchege (*Precision Radar Electro-Technology*, February 1965) to illustrate that the product of the RMS time delay error and RMS frequency error is

$$\sigma_{Te}\sigma_f = \frac{1}{B_e T_e (2E/N_o)} \tag{1.75}$$

and, when replaced by the radar uncertainty principle, yields

$$\sigma_{Te}\sigma_f \leq \frac{1}{\pi(2E/N_o)} \tag{1.76}$$

Time delay (range) and frequency (velocity) may be simultaneously measured to small error values by employing large $B_e T_e$ time bandwidth product waveforms such as pulse compression (linear FM) techniques.

The laser radar system's inherent capability to provide five-dimensional target data R, V, and I, angle-angle with optical resolution capability has resulted in laser radar systems being evaluated for precision fire control and autonomous missile guidance applications.

In these applications, efficient utilization of the average power of the laser transmitter is required in order to maximize the scan field of the sensor. Present systems use both high repetition rate pulse- and CW-based waveforms. In 1960, Cook [31] showed that the transmitter average power could be efficiently utilized through the use of pulse compression techniques. Here a CW-type source can be made to have the same range measurement qualities as a short-pulse system. Figure 1.19 shows the output of a CW source changing linearly from a frequency F_1 to a frequency F_2 in a time T_{FM}. The transmitted pulse (T_{FM}) upon reception is applied to a dispersive filter having a time delay linearly related to the inverse of frequency (i.e., higher frequencies have longer time delays than lower frequencies).

The filter allows the power at each frequency to be linearly delayed so that the power is compressed into a shorter time (τ) corresponding to that of a short-pulse system.

Similarly, the peak power (P_{PK}) at the filter output is increased such that

$$P_{PK} = P_{CW}\frac{T_{FM}}{\tau} \quad \text{or if} \quad B = 1/\tau = P_{CW}T_{FM}B \tag{1.77}$$

The CW power contained within the transmitted pulse is compressed by the time-bandwidth product to be equal to that of a peak power transmission in a pulse length (τ).

For example, if the transmitted power is 1W, the frequency modulation pulse length is 10 μs, and the short pulse length desired is 100 ns ($B = 1/\tau = 10$ MHz),

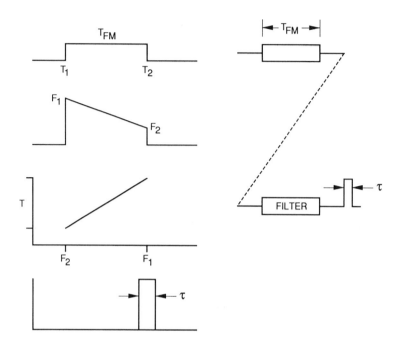

Figure 1.19 Pulse compression, ©IEEE. (Courtesy of IEEE.)

then the equivalent peak power is 100W (the CW power times the time bandwidth product). Achievement of this pulse compression ratio assumes that the system has a linear frequency deviation and that time compression is better than 0.1% in order to achieve the desired compression.

Figure 1.20 [30] illustrates some of the time and frequency expressions for a variety of waveform shapes.

Because of the very narrow pulse widths that can be obtained with lasers, the range errors may be small, as can be observed in Figure 1.21 for that of a 100-ns pulse length (τ). Many of the solid-state lasers can provide high peak power levels, and are therefore used as range finders.

The pulse and frequency-modulated waveforms have a velocity and range measurement accuracy capability as a function of receiver processor bandwidth, as shown in Figure 1.22. Here, it may be observed that there is a tradeoff required for pulse systems to simultaneously measure range and Doppler velocity (i.e., good range accuracy—poor velocity accuracy). Assuming a matched filter bandwidth of 1 MHz, the range accuracy for the pulse (1ms pulse length) and FM system would be identical (15 m), and the velocity accuracy would be 0.5 m/s for a 10.6-μm wavelength.

Improving the range accuracy for the system to 1.5m results in the pulse system requiring a bandwidth of 10 megahertz (100-ns pulse length) yielding a velocity

Waveform	Range Accuracy	Frequency Accuracy
Rectangular (τ) pulse where rise time is limited by IF bandwidth	Leading Edge Only $$\sigma_R = \frac{c}{2}\left(\frac{\tau}{2B_e E/N_o}\right)^{1/2}$$ Leading and Trailing Edge $$\sigma_R = \frac{c}{2}\left(\frac{\tau}{4B_e E/N_o}\right)^{1/2}$$	$$\sigma_f = \frac{\sqrt{3}}{\pi\tau(2E/N_o)^{1/2}}$$
Triangular Pulse where T_B is the pulsewidth at the base	$$\sigma_R = \frac{c}{2}\frac{T_B}{\sqrt{12}(2E/N_o)^{1/2}}$$	$$\sigma_f = \frac{(10)^{1/2}}{\pi T_B\left(\frac{2E}{N_o}\right)^{1/2}}$$
Gaussian Pulse	$$\sigma_R = \frac{c}{2}\frac{1.18}{\pi B(2E/N_o)^{1/2}}$$	$$\sigma_f = \frac{B}{1.18(2E/N_o)^{1/2}}$$
Pulse Compression	$$\sigma_R = \frac{c}{2}\left(\frac{\sqrt{3}}{\pi B(2E/N_o)^{1/2}}\right)$$	$$\sigma_f = \left(\frac{B}{4\tau E/N_o}\right)^{1/2}$$
Am Modulation	$$\sigma_{RAM} = \frac{c}{4\pi F_{AM} m \sqrt{SNR}}$$	
Aperture Illumination Rectangular	$$\sigma_\theta = \frac{0.63\theta_B}{(2E/N_o)^{1/2}}$$	

For cosine or triangular aperture illumination functions, the 0.63 factor changes to 0.85 and 0.9, respectively. θ_B is the one-half power beamwidth, σ_θ is based on one-way voltage gain (i.e., receive only): in a radar case (i.e., two-way) signal amplitude is proportional to the two-way voltage gain pattern.

where
σ_R = RMS range error σ_f = RMS frequency error
σ_{Te} = RMS time error $\sigma_R = \frac{c}{2}\sigma_{Te}$
E = signal energy σ_θ = RMS angular error
N_o = noise power per cycle F_{AM} = AM modulation frequency
    m = modulation index

Figure 1.20 Theoretical accuracy of radar measurements. (Courtesy of IEEE [30]).

measurement accuracy of 5 m/s while the pulse compression system requires a bandwidth of 100 kHz (for the chirp characteristic chosen) and results in a velocity accuracy of 5 cm/s.

This trade is not required for FM systems, because the slope of the measurement accuracy plots are in the same direction, and chirp characteristics may be chosen to result in good simultaneous range and velocity measurement accuracy.

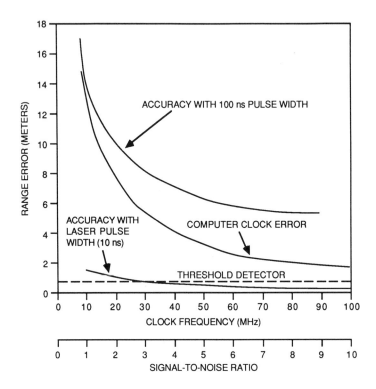

Figure 1.21 System range error *versus* signal-to-noise ratio.

The curves are plotted for SNR = 50 for the FM system and 100 for the pulse systems, and assumes the signals are at an intermediate frequency within the receiver such that the processing bandwidth is the inverse of the pulse width. For convenience, the FM chirp characteristic used is 10^{12} Hz/s.

Due to the short wavelength operation of laser radars, Doppler shifts are extremely large compared to microwave systems. This results in IF signals in the receiver allowing short measurement time intervals, thereby permitting simultaneous range, velocity, and angle measurements.

The pointing resolution [32,33] from a laser radar quadrant-photodiode detector is given by

$$\sigma_\theta = \frac{3\pi}{16} \frac{\lambda}{D} \frac{1}{\sqrt{\text{SNR}}} \frac{1}{S_T} \qquad (1.78)$$

where $\sqrt{(\text{SNR})}$ is equal to the voltage—SNR, and S_T is a complex function related

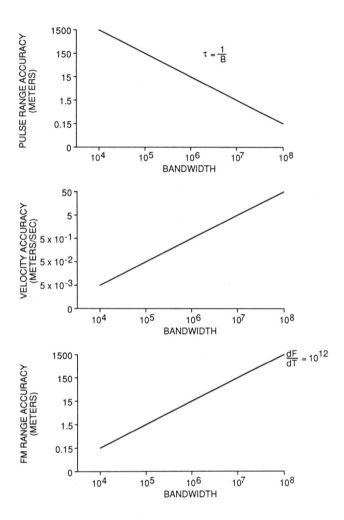

Figure 1.22 Range/velocity accuracy *versus* bandwidth.

to the image's spatial distribution and obscuration ratio. An unobscured aperture S_T is approximately equal to one. The geometrical factor, $3\pi/16$, comes from integration over the Airy disk.

By differencing consecutive angular position measurements over time, a *cross-range velocity* estimate can be made based on a linear-plus-quadratic-data-fit Kalman filter. The accuracy of the estimate is given by [34]

$$\sigma_{R_L} = R\sigma_\theta \sqrt{12\Delta T}\, T_0^{-3/2} \tag{1.79}$$

where

ΔT = interpulse time
T_0 = observation time

1.10 IMAGING SYSTEMS

1.10.1 Angle-Angle Imaging

Angle-angle imaging uses the receiver aperture resolving capability to image the target backscattered illumination onto the receiver detector(s), where the optical photon energy is then converted to electrons and processed to provide an image of the target field. The transmitter aperture is selected to provide the illumination format (i.e., pencil, floodlight, pushbroom beam) in order to optimally create the required image.

1.10.1.1 Resolved Beam Imaging

The range performance of an extended target, coherent, angle-angle, pencil beam imaging system, where the transmitter and receiver have matched fields of view may be expressed as

$$R = \left(\frac{\eta_{ATM} \, \pi^2 \, \eta_{SYS} \, E_T \, \rho \, D_R^2 \, \lambda}{16 \, \text{SNR} \, hC} \right)^{1/2} \qquad (1.80)$$

Parametric sensitivity of this expression shows that the extended target range performance is linearly related to the optics diameter, to the square root related to the energy, and to the wavelength.

Aperture size limitations and imaging angular resolution requirements typically drive this approach to shorter wavelengths for small targets. Additionally, scan field requirements result in high pulse repetition rates for matched field systems. In order to alleviate the scan field pulse repetition frequency (PRF) difficulties, the transmitter aperture is typically reduced in either one dimension (fan beam) or two dimensions (floodlight beam).

For these conditions, the transmitter energy requirement increases to maintain the pixel illumination energy level, and the number of detectors increases accordingly; that is, assuming a 10m × 10m field and a pixel resolution requirement of 1 m × 1 m, the receiver would have 10 detectors in a line for a scanning fan beam and 100 detectors in a 10 × 10 array for a floodlight beam.

1.10.1.2 Floodlight Field Imaging

For a floodlight configuration, the transmitter energy is distributed over a spot size larger than the resolution of a pixel, and the range equation becomes

$$R = \left[\frac{E_T \rho d A \lambda^2 \, \eta_{SYS} \, \eta_{ATM} \, K_a^2}{4 P_R \theta_T^2 \theta_R^2} \right]^{1/4} \quad (1.81)$$

Typically the pointing and tracking system accuracy determines the size of the illumination field (θ_T), and the imaging resolution requirement determines the receiver angular resolution (θ_R). At long ranges, usually large received aperture size is required at long wavelengths, to obtain the resolution requirements, and results in shorter wavelength operation in the floodlighting mode.

1.10.2 Range-Doppler Imaging

The use of the Doppler properties of the received signal allows Doppler beam sharpening techniques to be used to improve the cross-range measurement capability of a given aperture, and alleviates the diffraction-limited aperture requirement to spatially resolve a target by angular resolving properties. This technology is well known in the microwave radar art and is known as *synthetic aperture radar* (SAR) technology. Here, the Doppler spectrum of a stationary target, measured from a moving platform, is used to sharpen the angular resolution of the antenna. In the case where the target is moving and spinning, the target spin characteristics are used to improve the angular resolution of the aperture. In this case, the related radar technology is referred to as *equivalent inverse synthetic aperture radar* (ISAR) technology.

Doppler Shift. The equation for the Doppler-shifted backscatter signal may be expressed as

$$f_d = \frac{2V}{\lambda} \cos\theta \quad (1.82)$$

where

V = target velocity
θ = angle between target velocity vector and line of sight

Because of the shorter operating wavelength associated with lasers, there is greater Doppler shift per meter per second of velocity.

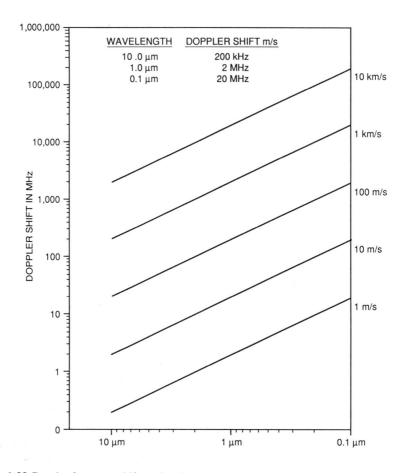

Figure 1.23 Doppler frequency shift produced by moving targets *versus* wavelengths.

Figure 1.23 illustrates the approximate Doppler frequency shift of targets having radial velocities ranging from 1 m/s to 10 km/s over a wavelength range from 0.1 to 10 µm.

A 3-cm wavelength radar would have a Doppler shift on the order of 67 Hz/m/s of target motion. Correspondingly, operation at 1 µm would yield 2 MHz of Doppler shift for every meter per second of target motion. The best sources of coherent laser radiation for Doppler detection occur at approximately 10 µm; at this wavelength, Doppler shift is approximately 200,000 Hz/m/s of target velocity.

In microwaves, if it is desirable to extract a Doppler shift of a target, that Doppler shift typically is masked within the sin x/x response of the transmitted pulse width and, as a result, Nyquist PRF sampling is used to sample the small shift. If the target had a 100-Hz Doppler, approximately 0.01 sec would be required to assess its velocity. The by-product of this situation is that the usual microwave system is of the high repetition rate pulsed Doppler variety. At optical wavebands, the Doppler shifts are sufficiently large for instantaneous measurement to be performed on a single pulse; therefore, low PRF systems can be configured.

1.10.3 Range-Velocity Ambiguity

As the pulse repetition rate increases, the distance over which the pulse provides a unique range measurement becomes shorter. Assuming the transmitter energy has the capability to reach a target at a multiple of the PRF range interval (R_{UNAMB}), a signal can exist in the receiver, which will be at a multiple of the unique range interval. This effect also occurs in velocity space. If the applied waveform is also simultaneously trying to measure velocity, a range-velocity ambiguity exists. The range ambiguity may be expressed as

$$R_{UNAMB} = \frac{CT_s}{2} \tag{1.83}$$

Similarly, the velocity ambiguity is

$$V_{UNAMB} = \frac{\lambda}{2T_s} \tag{1.84}$$

where T_s is the interpulse spacing. The ability to create a coherent image assumes that the down-range and cross-range measurements are correlated; this is distinctly different from independent range and velocity target measurement, which are not time correlated. The time correlation associated with the selected waveform allows a unique range and cross-range measurement to be attributed to the target, and can be used to construct a target image.

The down-range and cross-range resolution may be expressed as

$$R_{down\ range}(R_D) = \frac{K_5 C}{2B} \tag{1.85}$$

$$R_{cross\ range}(R_C) = \frac{K_6 \lambda}{2 \sin\theta_c \cos\Phi} \tag{1.86}$$

where

B	=	bandwidth
λ	=	wavelength
C	=	speed of light
θ_C	=	angle through which object turns in processing time
Φ	=	angle between the target spin vector and its position on a plane normal to the line of sight to the target (aspect angle)
K_5, K_6	=	broadening factor

The cross-range equation determines the pulse burst length, assuming a waveform matched to the target coherence time. It should be pointed out that the target coherence time will be longer for discrete scatterers than for a diffuse target. This is illustrated in Figure 1.24, from S. Weiner, MIT Lincoln Laboratory [35]. An example of diffuse spinning target waveform considerations for a cylinder are illustrated in Figure 1.25. This example illustrates the relationships between the down-range and cross-range resolution. The down-range resolution of 20 cm is obtained by transmitting a pulse of ~1 ns or an equivalent 1-GHz frequency-based waveform. Assuming a 1m diameter cylinder, rotating at 1 revolution per second, the Doppler

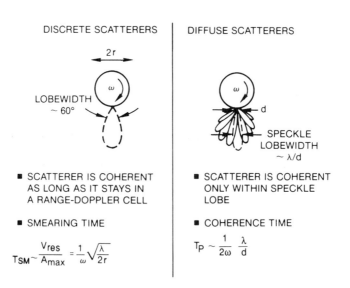

Figure 1.24 Target coherent observation time limitations [35].

A SYSTEM HAVING A RANGE RESOLUTION OF 20 cm (FROM MATCHED FILTER THEORY) REQUIRES:

- TIME DOMAIN = 1 ns PULSE
- FREQUENCY DOMAIN = 1 GHz MODULATION BANDWIDTH

DETECTION

COHERENT INTEGRATION TIME IS DEPENDENT UPON TARGET DECORRELATION TIME IF:

$$V = \frac{Wr}{T}$$

IF $\quad r = 0.5 \text{ m} \quad w = 2\pi/T$

AND $\quad f_{d_{MAX}} = \frac{2V}{\lambda} = \frac{2 \times 2\pi \times 0.5}{10^{-5} \text{ m}} = 0.6 \text{ MHz}$

TOTAL DOPPLER = $2f_{d_{MAX}}$ = 1.2 MHz OR

TARGET CORRELATION TIME (DETECTION) 0.8 μs @ 10 μ

0.08 μs @ 1 μ

0.008 μs @ 0.1 μ

IMAGING

IF 1.2 MHz CORRESPONDS TO 1 m DIAMETER, THEN 20 cm CORRESPONDS TO:

240 kHz @ 10 μ i.e. τ = 4 μs

OR 2.4 MHz @ 1 μ i.e. τ = 0.4 μs

OR 24 MHz @ 0.1 μ i.e. τ = 0.04 μs

Figure 1.25 Waveform consideration.

excursion over the cylinder is 1.2 MHz, corresponding to the 1m diameter. A Doppler resolution of one-fifth of 1.2 MHz or 240 kHz then corresponds to a 20-cm segment of the target. The image coherence time for this resolution is approximately 4 μs and represents the length of the pulse burst (T_p). The interpulse spacing (T_s) within the 4-μs pulse burst of 1-μs pulses is determined by the range-velocity ambiguity relationship illustrated in Figures 1.26 and 1.27. The pulse repetition rate of the 1-μs pulses must meet the Nyquist sampling criteria and provide at least two samples of the Maximum Doppler waveform (see Figure 1.27).

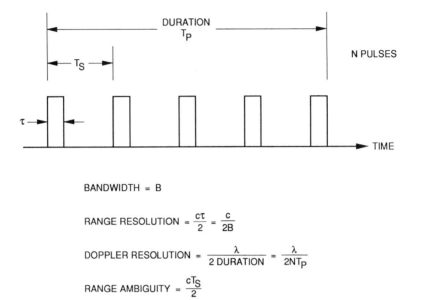

BANDWIDTH = B

RANGE RESOLUTION = $\dfrac{c\tau}{2} = \dfrac{c}{2B}$

DOPPLER RESOLUTION = $\dfrac{\lambda}{2\,\text{DURATION}} = \dfrac{\lambda}{2NT_P}$

RANGE AMBIGUITY = $\dfrac{cT_S}{2}$

DOPPLER AMBIGUITY = $\dfrac{\lambda}{2T_S}$

Figure 1.26 Burst parameters.

Table 1.2
Waveform Performance Parameters (Courtesy of MIT Lincoln Laboratory [36].)

			Performance Issues	
Waveform and Parameters	Measurement	Resolution	Ambiguity Peaks	Side-lobe to Main-lobe Ratio
Unmodulated Pulse	Range	$cT/2$	None	0
(T)	Doppler	$1/T$	None	-13 dB
Unmodulated Pulse	Range	$cT/2$	$cT_s/2$	0
Train (T, N_pT_s)	Doppler	$1/(N_pT_s)$	$1/T_s$	-13 dB
LFM Chirp Pulse	Range	$c/2B$	None	-13 dB
(T, B)	Doppler	$1/T$	None	-13 dB
LFM Chirp Pulse Train	Range	$c/2B$	$cT_s/2$	-13 dB
(T_p, B, N_pT_s)	Doppler	$1/(N_pT_s)$	$1/T_s$	-13 dB

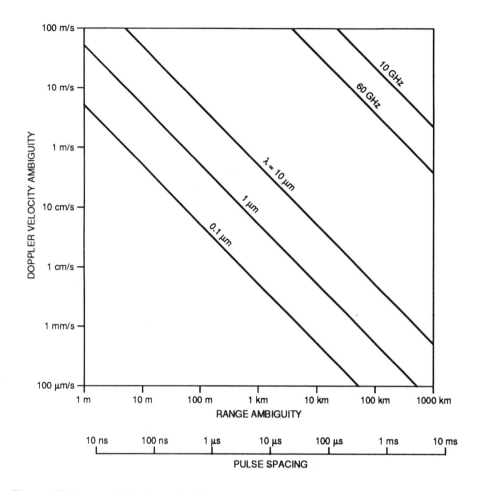

Figure 1.27 Range and Doppler ambiguities.

For a discrete scatterer the maximum Doppler frequency is dependent upon the time for a point scatterer to rotate 360°, and results in a substantially longer coherence time (T_{sm}). Using the earlier diffuse target conditions, the image coherence time becomes 500 μs.

Kachelmyer ("Range-Doppler Imaging with a Laser Radar") [36–42] illustrated the waveform range-Doppler ambiguity function in the complex domain. Figure 1.28 illustrates two waveforms (pulse and linear frequency modulated chirp) where

T_p = single-pulse burst duration
T_s = interpulse spacing

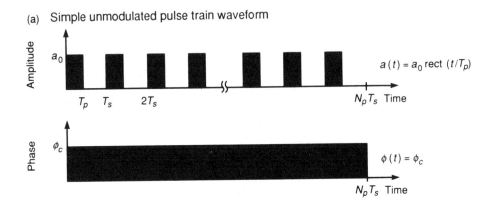

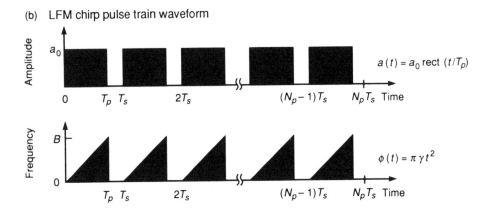

Figure 1.28 Two basic waveforms. (Courtesy of MIT Lincoln Laboratory [36].)

T_B = time bandwidth product
N_P = number of pulses in biphase shift keyed pulse train

and Figure 1.29 shows the range-Doppler ambiguity function in the complex domain. Table 1.2 illustrates the waveform capabilities of providing resolution and side-lobe levels.

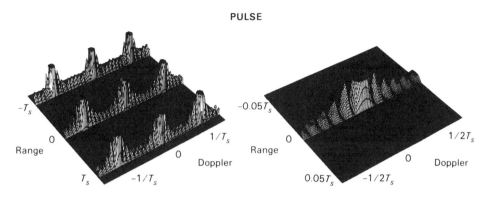
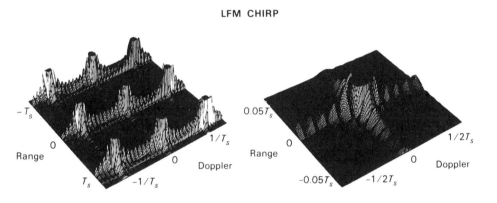

Figure 1.29 Range-Doppler ambiguity functions for a 12-pulse train for two waveforms. Each waveform has a single-pulse TB product of 127, and a corresponding pulse train TB product of 1524. Responses are truncated at 20% of the mainlobe peak value. (Courtesy of MIT Lincoln Laboratory [36].)

Chapter 2
Atmospheric Propagation

Accurate propagation constants for lasers are a sensitive function of wavelength and atmospheric conditions. Useful lasers operate at wavelengths in the ultraviolet, visible, near-infrared, mid-infrared, and far-infrared regions where atmospheric windows occur. Figure 2.1 illustrates the transmission of over 1000 ft of the atmosphere at sea level [43]. Approaching the electromagnetic spectrum from the radar wavebands, the first atmospheric window that allows propagation to occur is the 8- to 14-μm band. Continuing past that point, the atmosphere is highly nontransmissive until the mid-infrared waveband between 3 and 5 μm. The absorption that occurs in the atmosphere, preventing transmission at the far- and mid-infrared bands, is caused basically by the absorption of atmospheric constituents such as ozone, water vapor, and carbon dioxide. Proceeding along, the next atmospheric windows occur in the near-IR region from 0.7 to 2.5 μm and, subsequently, the visible region. When these smaller wavebands are being used, haze and scattering attenuation caused by particulates in the atmosphere are of prime concern.

Table 2.1 illustrates the amount of attenuation as a function of wavelength for both laser and microwave wavebands. The chart is compiled from broadband data and does not include the fine-grain atmospheric absorption values into which laser lines may drift occasionally (and which can be prevented by suitable transmitter design).

In the visible waveband where the ruby laser operates (0.69 μm) the atmosphere "seeing" conditions have a very dramatic effect upon the amount of attenuation that the laser beam encounters. Visibility on a clear day is 15 km, with a corresponding one-way attenuation of 1 dB/km. Introduction of haze into the atmosphere reducing the visibility to 3 km increases the attenuation to 5.2 dB/km. Moving to the 1.06-μm wavelength (that associated with a YAG laser), it is seen from the table that these conditions produce less attenuation than they did in the visible region; further movement to the mid-infrared at 3.84 μm (the wavelength of a strong emission line of the DF chemical laser) results in only 2.3 dB/km of attenuation in a haze. Progressing into the far-infrared at 10.6 μm decreases the haze

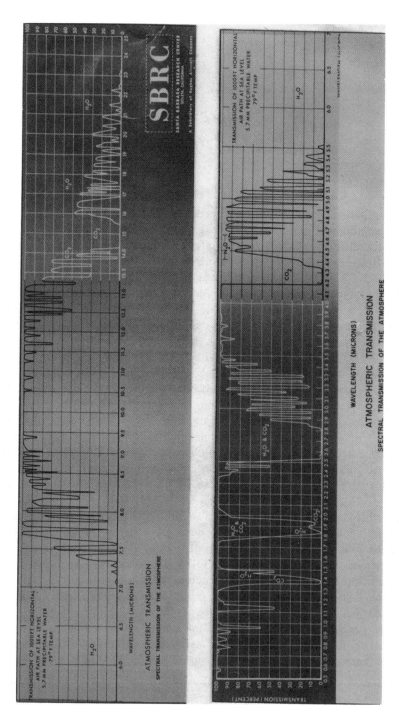

Figure 2.1 Atmospheric transmission for 1000-ft (0.3-km) path at sea level for 0.5 to 25 μm.

attenuation to 1 dB/km; the predominant attenuation mechanism in this region is absorption by H_2O. The second column under the 10.6-μm wavelength addresses the problems of humidity in subarctic, midlatitude, and tropical environments. McClatchey's [44] data for the atmospheric attenuation under tropical environments result in a factor of 2.8 dB/km; correspondingly, a chief concern is system operation in fog and rain. The third column under $\lambda = 10.6$ μm illustrates specific data accumulated by Goodwin [45], Rensch [46], and Chu [47]. From this listing it is clear that light fog, having a visibility in the visible waveband of 2 km, would yield an attenuation of approximately 1 dB/km. A radiation fog having visibility of 0.5 km would increase the attenuation to about 2.3 dB/km. These figures may be contrasted with those commonly found in microwave radar textbooks (Barton [48] and Skolnik [49]) for attenuation at microwave wavelengths from 3 cm to 0.3 cm. These last three columns indicate the ability of longer wavelength electromagnetic systems to penetrate fog with less attenuation.

The data with regard to rain are not necessarily so apparent. The table indicates that a rainfall rate of 12.5 mm/hr has equivalent attenuation from 0.7 μm to 0.3 cm. Subsequent columns in this chart show the wavelength attenuation at 1-cm (2.8 dB/km) and 3-cm (0.2 dB/km) systems in 12.5 mm/hr of rain.

Atmospheric attenuation coefficient at sea level is composed of Rayleigh, ozone, and aerosol scattering σ_s as shown in Figure 2.2, where the atmospheric transmission (T_a) may be obtained by $T_a = e^{-\sigma_s L}$, where L is the atmospheric path length. Figure 2.3 illustrates the same sea-level coefficient when related to atmospheric "seeing" conditions.

Attenuation coefficients covering the region from 2.9 through 10.8 μm, matching the transmitting wavelength of the HF/DF, CO, and CO_2 laser lines, are shown in Table 2.2 from *The Infrared Handbook* by Wolfe and Zissis [15]. The data in the table were derived from R.A. McClatchey and J.E.A. Selby [50], and are shown in terms of wave number ($1/\lambda$) in inverse centimeters.

Figures 2.4 and 2.5 illustrate the atmospheric transmission at 10.59 μm due to water vapor and carbon dioxide and water vapor, respectively [56]. Figure 2.6 relates the visibility condition in the visible to the liquid H_2O content of the atmosphere [57]. Atmospheric attenuation in turn is related to visibility via experimental data. Two equations derived by DiMarzio [60] in Figure 2.7 illustrate the approximate curve data fit.

Rain attenuation theory and experimental data are shown in Figure 2.8 [58–61]. In this curve, the attenuation coefficient is related to rain rate. Snow attenuation theory and experimental data are shown in Figure 2.9 [62].

Computer models such as Lowtran and Fascode may be used to calculate specific atmospheric transmissivity. However, atmospheric transmission can be approximately characterized by the curves shown in Figures 2.10 and 2.11. Accurate Fascode values for specific lasers for the case of the vertical propagation to space from 1.2 km altitude are shown in Table 2.3 [33].

Table 2.1
Attenuation (One-Way) *Versus* Wavelength

dB/km	dB/nmi	0.7 μm	1.06 μm	3.8 μm	10.6 μm	0.3 cm [48,49]	1 cm [48,49]	3 cm [48,49]
0.2	0.4	Extremely clear	Extremely clear	Clear ($V_1 = 23$ km)	Subarctic winter			12.5 mm/hr
0.5	0.9	Standard clear ($V_1 = 25$ km)*	Standard clear		Clear		2.5 mm/hr	
0.8	1.5		Clear			Rain 2.5 mm/hr	Fog (advection) ($V_1 = 100$ ft)	
1	1.8	Clear ($V_1 = 15$ km)		Haze $V_1 = 5$ km	Haze subarctic summer	Light fog** [45] ($V_1 = 2$ km)		50 mm/hr
1.5	2.8	Light haze ($V_1 = 8$ km)	Light haze		Midlatitude summer	Light rain [46] 2.5 mm/hr		
1.7	3.1					Fog** [45] ($V_1 = 0.5$ km)	(advection) ($V_1 = 400$ ft)	
2.3	4.2		Medium haze	Haze $V_1 = 3$ km				

2.8	5.2	Medium haze ($V_1 = 5$ km)	Tropical [44]	12.5 mm/hr
3.9	7.1			
5.2	9.5	Haze ($V_1 = 3$ km)		2.5 mm/hr
5.5	11.0	Medium rain 12.5 mm/hr	Medium rain [46,47] 12.5 mm/hr	25 mm/hr
9	16.5	Light fog $V_1 = 2$ km	Fog (advection) ($V_1 = 100$ ft)	30 mm/hr

*Visibility

**Fog is usually characterized by its optical visibilty. However, this description is inadequate in that the same visibility conditions can arise for fog of different liquid water content and/or different water droplet size distribution. It is the liquid water content alone that determines fog's attenuation at microwave frequencies, with the attenuation directly proportional to the liquid water content. Hence, fogs with the same visibility but with different particle size distributions and liquid content can produce different attenuations at microwave frequencies. Advection fogs are typically formed over open water as a result of the advection of warm, moist air over colder water. (These fogs have a higher moisture content than radiation fogs having the same visibility.) Radiation fogs—pure radiation fog is that which forms in air that has been over land during the daylight hours preceeding the night of its formation. An advection fog can have five times the liquid water content of a radiation fog with the same visibility conditions. This fact adds complexity to the problem of comparing optical and microwave frequency fog attenuation.

Sources: *Compendium of Meteorology*, H.R. Byers, H.E. Landsberg, H. Wexler, B. Haurwitz, A.F. Spilhaus, H.C. Willett, H.G. Houghton, American Meteorological Society, Boston, Mass., 1951, Waverly Press Inc. Baltimore, Maryland, pp. 1179–1189. George, Joseph, *Fog*, Eastern Airlines, Atlanta, GA.

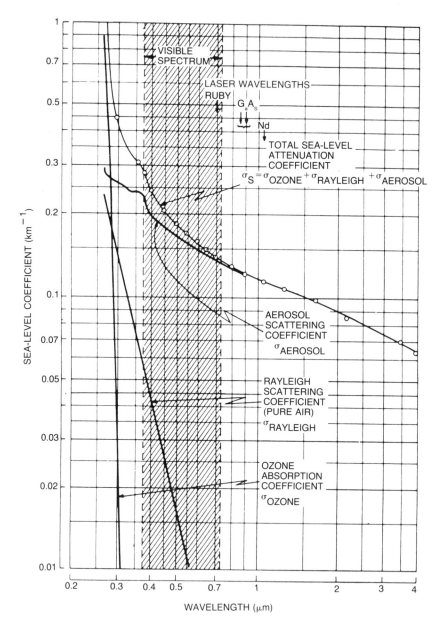

Figure 2.2 Calculated atmospheric attenuation coefficients for horizontal transmission at sea level in model clear standard atmosphere. (Courtesy of RCA [4].)

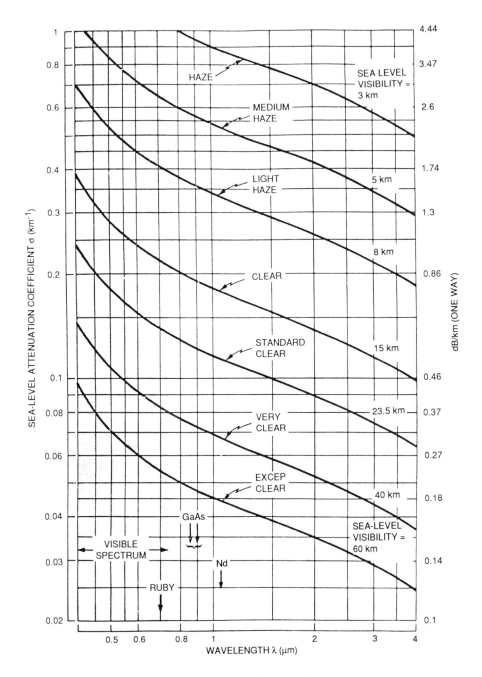

Figure 2.3 Approximate variation of attenuation coefficients with wavelength at sea level for various atmospheric conditions (neglects absorption by water vapor and carbon dioxide). (Courtesy of RCA [4].)

Table 2.2
Attenuation Coefficients for Laser Frequencies [50]

	CO Laser Parameters			Atmospheric Absorption Coefficients (km^{-1})			
				$z = 0$ km, Sea Level			$z = 12$ km
	Band	Rot. ID	$v(cm^{-1})$	k_{trop}	k_{mw}	k_{sw}	k_{mw}
(a)	1-0	P2	2135.549	0.661	0.249	0.224	0.266
		P14	2086.325	0.409	0.202	0.176	0.141
		P17	2073.267	0.608	0.159	0.104	0.0511
		P18	2068.849	0.268	0.101	0.0792	0.0352
		P21	2055.402	0.141	0.0750	0.0654	0.0112
		P22	2050.856	0.152	0.0522	0.0392	0.00630
		P25	2037.027	0.411	0.0765	0.0369	0.00574
		P26	2032.354	0.178	0.0292	0.0124	0.000813
		P27	2027.651	0.757	0.137	0.0477	0.000650
		P30	2013.353	0.548	0.0784	0.0230	0.000077
(a)	2-1	P1	2112.977	0.0935	0.0144	0.00665	0.00035
		P2	2109.132	0.0525	0.0168	0.0126	0.00902
		P3	2105.256	0.120	0.0264	0.0125	0.0038
		P4	2101.342	0.122	0.0246	0.0127	0.0055
		P7	2089.393	1.52	0.191	0.0527	0.00671
		P8	2085.343	0.186	0.0346	0.0218	0.00196
		P9	2081.258	0.151	0.0276	0.0140	0.00109
		P11	2072.987	0.366	0.0733	0.0332	0.00268
		P12	2068.802	0.240	0.0761	0.0563	0.00427
		P15	2056.046	0.144	0.0218	0.0118	0.000605
		P16	2051.729	1.09	0.0846	0.0283	0.000769
		P17	2047.379	0.350	0.0718	0.0413	0.00118
		P19	2038.582	0.365	0.0542	0.0190	0.000178
		P21	2029.656	0.213	0.0314	0.00956	0.000032
		P22	2025.145	0.537	0.0746	0.0221	0.000079
		P25	2011.423	0.407	0.0577	0.0167	0.000014
		P26	2006.786	0.801	0.108	0.0300	0.000020
		P27	2002.118	0.320	0.0504	0.0156	0.000016
		P28	1997.419	0.938	0.157	0.0501	0.000045
	3-2	P1	2086.594	0.479	0.0565	0.0263	0.00305
		P2	2082.784	0.114	0.0181	0.00920	0.00084
		P3	2078.940	0.630	0.171	0.125	0.045
		P4	2075.061	0.333	0.0558	0.0216	0.0064
		P5	2071.148	0.123	0.0235	0.0125	0.000861
		P6	2067.200	0.679	0.122	0.0508	0.00181
		P7	2063.218	0.801	0.130	0.0561	0.00152
		P8	2059.203	0.571	0.0937	0.0365	0.000655
		P10	2051.071	0.414	0.0581	0.0236	0.000598
		P11	2046.954	0.851	0.104	0.0292	0.000119
		P12	2042.804	1.49	0.225	0.0735	0.000429

Table 2.2 (continued)

	CO Laser Parameters			Atmospheric Absorption Coefficients (km^{-1})			
				z = 0 km, Sea Level			z = 12 km
	Band	Rot. ID	$v(cm^{-1})$	k_{trop}	k_{mw}	k_{sw}	k_{mw}
		P13	2038.621	0.367	0.0525	0.0174	0.000122
		P14	2034.405	0.882	0.0896	0.0217	0.000239
		P15	2030.157	0.317	0.0406	0.0116	0.000073
		P16	2025.875	1.13	0.166	0.0513	0.000365
		P17	2021.561	0.734	0.098	0.0277	0.000066
		P19	2012.835	0.739	0.102	0.0290	0.000077
		P20	2008.424	1.68	0.231	0.0654	0.000044
		P21	2003.981	0.299	0.0416	0.0127	0.000117
		P25	1985.891	1.06	0.155	0.0455	0.000030
		P26	1981.290	0.843	0.0773	0.0188	0.000011
		P27	1976.658	1.15	0.214	0.0735	0.000094
		P28	1971.995	0.607	0.0944	0.0290	0.000040
		P30	1962.577	1.37	0.216	0.0660	0.000058
(a)	4-3	P2	2056.506	0.127	0.0568	0.0497	0.00233
		P3	2052.697	0.0955	0.0198	0.0114	0.000392
		P4	2048.853	0.283	0.0616	0.0406	0.00151
		P5	2044.975	0.779	0.125	0.0407	0.000133
		P7	2037.116	0.568	0.0802	0.0305	0.00110
		P8	2033.135	0.172	0.0215	0.00596	0.000012
		P9	2029.121	0.180	0.0284	0.00939	0.000049
		P10	2025.074	0.503	0.0708	0.0214	0.000069
		P11	2020.993	0.859	0.119	0.0338	0.000050
		P13	2012.731	0.581	0.0816	0.0234	0.000022
		P14	2008.550	1.43	0.203	0.0590	0.000053
		P15	2004.337	0.302	0.0406	0.0117	0.000001
		P17	1995.812	1.12	0.170	0.0513	0.000039
		P20	1982.783	0.507	0.0753	0.0225	0.000017
		P21	1978.375	0.281	0.0446	0.0141	0.000048
		P22	1973.936	0.386	0.0607	0.0187	0.000016
(a)	5-4	P2	2030.297	0.186	0.0236	0.00682	0.000011
		P6	2014.993	1.62	0.229	0.0666	0.000117
		P7	2011.082	1.02	0.138	0.0392	0.000120
		P8	2007.137	1.70	0.225	0.0623	0.000219
		P9	2003.158	0.373	0.0502	0.0144	0.000018
		P11	1995.100	1.61	0.243	0.0731	0.000075
		P14	1982.764	0.496	0.0730	0.0217	0.000017
		P15	1978.586	0.266	0.0416	0.0129	0.000016
		P16	1974.376	0.412	0.0631	0.0194	0.000016
		P21	1952.838	0.900	0.145	0.0453	0.000046
		P25	1935.035	1.29	0.205	0.0681	0.001500
		P26	1930.506	1.13	0.180	0.0563	0.000071

Table 2.2 (continued)

	CO Laser Parameters			Atmospheric Absorption Coefficients (km^{-1})			
				z = 0 km, Sea Level			z = 12 km
	Band	Rot. ID	$v(cm^{-1})$	k_{trop}	k_{mw}	k_{sw}	k_{mw}
(a)	6-5	P2	2004.155	0.588	0.0587	0.0151	0.000026
		P3	2000.415	0.783	0.134	0.0434	0.000040
		P4	1996.641	1.089	0.155	0.0464	0.00039
		P7	1985.115	0.738	0.108	0.0319	0.000024
		P8	1981.205	1.55	0.119	0.0257	0.000013
		P9	1977.261	0.437	0.0737	0.0238	0.000023
		P10	1973.284	0.432	0.0669	0.0205	0.000022
		P15	1952.901	0.917	0.147	0.0459	0.000044
(b)	7-6	P3	1974.409	0.424	0.0641	0.0196	0.000016
		P4	1970.670	1.16	0.176	0.0529	0.000042
		P6	1963.089	1.26	0.195	0.0594	0.000052
		P7	1959.247	0.969	0.152	0.0469	0.000048
		P14	1931.380	1.36	0.212	0.0653	0.000106
(c)	1-0	P11	3436.12	2.21	0.221	0.0542	0.0000287
(c)		P12	3381.50	0.496	0.0751	0.0231	0.000022
(c)	2-1	P8	3435.17	2.01	0.209	0.0512	0.0000267
(c)	3-2	P6	3373.46	0.364	0.0537	0.0168	0.000029
(d)	4-3	P8	3130.09	0.801	0.148	0.0554	0.000295
		P9	3083.83	1.12	0.211	0.0808	0.000806
(d)	5-4	P4	3150.67	0.498	0.126	0.0736	0.00229
(d)	6-5	P6	2921.74	0.586	0.0453	0.0103	0.000077
		P7	2880.70	0.0430	0.00424	0.00121	0.000006
		P8	2838.59	0.369	0.0654	0.0218	0.000044
(e)	1-0	P1	2884.934	0.414	0.123	0.0772	0.00316
(e)		P2	2862.652	0.0540	0.0115	0.00485	0.00316
(e)		P3	2839.779	0.0386	0.00725	0.00266	0.000038
(e)		P4	2816.362	0.0837	0.0190	0.0104	0.00108
(d)		P11	2003.56	0.367	0.0480	0.0138	0.000085
(d)		P12	1981.38	0.476	0.0557	0.0152	0.000010
		P40	924.970	0.514	0.0359	0.0112	0.000812
		P38	927.004	0.521	0.0423	0.0154	0.00164
		P36	929.013	0.744	0.0581	0.0190	0.00211
		P34	930.997	0.538	0.0536	0.0227	0.00311
		P32	932.956	0.557	0.0650	0.0302	0.00520
		P30	934.890	0.572	0.0737	0.0360	0.00677
		P28	936.800	0.588	0.0852	0.0440	0.00887
		P26	938.684	0.583	0.0853	0.0447	0.00955
		P24	940.544	0.603	0.0955	0.0517	0.0118
		P22	942.380	0.606	0.1021	0.0569	0.0136
		P20	944.190	0.609	0.0958	0.0521	0.0125
		P18	945.976	0.635	0.1223	0.0717	0.0186

Table 2.2 (continued)

	CO Laser Parameters			Atmospheric Absorption Coefficients (km^{-1})			
				z = 0 km, Sea Level			z = 12 km
	Band	Rot. ID	v(cm^{-1})	k$_{trop}$	k$_{mw}$	k$_{sw}$	k$_{mw}$
		P16	947.738	0.572	0.0747	0.0378	0.00897
		P14	949.476	0.607	0.1101	0.0642	0.0173
		P12	951.189	0.591	0.1058	0.0619	0.0171
		P10	952.877	0.596	0.1008	0.0580	0.0161
		P8	954.541	0.553	0.0817	0.0452	0.0123
		P6	956.181	0.513	0.0615	0.0314	0.00810
		P4	957.797	0.484	0.0498	0.0236	0.00573
		P2	959.388	0.978	0.0753	0.0282	0.00609
		R0	961.729	0.456	0.0347	0.0130	0.00234
		R2	963.260	0.461	0.0401	0.0170	0.00367
		R4	964.765	0.478	0.0502	0.0241	0.00590
		R6	966.247	0.519	0.0614	0.0308	0.00783
		R8	967.704	0.505	0.0663	0.0352	0.00931
		R10	969.136	0.510	0.0714	0.0389	0.0104
		R12	970.544	0.578	0.0788	0.0418	0.0109
		R14	971.927	0.556	0.0796	0.0427	0.0110
		R16	973.285	0.554	0.0799	0.0425	0.0106
		R18	974.618	0.522	0.0755	0.0405	0.0101
		R20	975.927	0.194	0.2140	0.0740	0.0109
		R22	977.210	0.674	0.0871	0.0398	0.00803
		R24	978.468	0.503	0.0641	0.0318	0.00699
		R26	979.701	0.484	0.0579	0.0280	0.00585
		R28	980.909	0.474	0.0529	0.0245	0.00471
		R30	982.091	0.552	0.0587	0.0240	0.00378
		R32	983.248	0.454	0.0436	0.0183	0.00324
		R34	984.379	0.455	0.0439	0.0158	0.00229
		R36	985.484	0.436	0.0357	0.0133	0.00176
		R38	986.563	0.428	0.0328	0.0114	0.00138
		R40	987.616	0.423	0.0306	0.0102	0.00121
(e)		P5	2792.437	0.0471	0.0106	0.00496	0.000157
(e)		P6	2767.914	0.0719	0.0184	0.00952	0.000672
(e)		P7	2743.028	0.0352	0.00801	0.00352	0.000043
(e)		P8	2717.536	0.114	0.0204	0.00718	0.000034
(e)		P9	2691.409	0.0248	0.00485	0.00252	0.000053
(e)		P10	2665.20	0.0237	0.00752	0.00489	0.000307
(e)		P11	2638.396	0.337	0.0664	0.0247	0.000187
(e)		P12	2611.125	0.0133	0.00394	0.00302	0.000090
(f)		P13	2584.91	0.0145	0.0102	0.00981	0.00390
(f)		P14	2557.09	0.0176	0.0180	0.0185	0.00335
(c)		P15	2527.06	0.0145	0.0155	0.0161	0.000565
(c)		P16	2498.02	0.0261	0.0282	0.0295	0.00103

Table 2.2 (continued)

	CO Laser Parameters			Atmospheric Absorption Coefficients (km^{-1})			
				z = 0 km, Sea Level			z = 12 km
	Band	Rot. ID	v(cm^{-1})	k_{trop}	k_{mw}	k_{sw}	k_{mw}
(c)	2-1	P3	2750.05	0.0401	0.00898	0.00403	0.000074
(c)		P4	2727.38	0.0378	0.00653	0.00272	0.000033
(c)		P5	2703.98	0.00528	0.00171	0.00118	0.0000307
(c)		P6	2680.28	0.0600	0.0139	0.00611	0.000069
(c)		P7	2655.97	0.0535	0.0134	0.00667	0.000733
(c)		P8	2631.09	0.00950	0.00348	0.00293	0.000761
(c)		P9	2605.87	0.0311	0.00776	0.00455	0.000110
(c)		P10	2580.16	0.282	0.0295	0.0311	0.00180
(c)		P11	2553.97	0.0144	0.0163	0.0177	0.000883
(c)		P12	2527.47	0.0140	0.0152	0.0158	0.000554
(c)		P13	2500.32	0.0240	0.0265	0.0278	0.000072
(c)		P16	2417.27	0.0811	0.0901	0.0943	0.00330
(c)	3-2	P3	2662.17	0.354	0.00790	0.00361	0.000047
(c)		P4	2640.04	0.0437	0.00914	0.00424	0.000075
(c)		P5	2617.41	0.00490	0.00276	0.00253	0.000090
(c)		P6	2594.23	0.0118	0.00557	0.00480	0.000152
(c)		P7	2570.51	0.0507	0.0560	0.0613	0.00557
(c)		P8	2546.37	0.0322	0.0356	0.0379	0.00228
(c)		P9	2521.81	0.0150	0.0164	0.0171	0.00599
(c)		P10	2496.61	0.0319	0.0298	0.0307	0.00107
(c)		P11	2471.34	0.0509	0.0491	0.0508	0.00184
(c)		P12	2445.29	0.0659	0.0728	0.0756	0.00266
(c)		P13	2419.02	0.0797	0.0885	0.0927	0.00325
(c)		P14	2392.46	0.141	0.199	0.115	0.00369
(c)	4-3	P5	2532.50	0.0134	0.0143	0.0148	0.000528
(c)		P6	2500.86	0.0199	0.0218	0.0228	0.000795
(c)		P7	2486.83	0.0318	0.0349	0.0356	0.00129
(d)		P8	2463.25	0.0681	0.0563	0.0571	0.00198
(d)		P9	2439.29	0.0686	0.0758	0.0794	0.00279
(d)		P10	2414.89	0.0829	0.0921	0.0964	0.00338
(d)	5-4	P7	2404.63	0.0878	0.0965	0.101	0.00354
	7-6	P8	2222.68	0.251	0.233	0.226	0.0102
		P10	2177.99	0.123	0.0979	0.0867	0.00297
		P11	2155.03	0.186	0.0344	0.0225	0.000846
		P12	2131.68	0.272	0.187	0.195	0.0311
(d)	8-7	P7	2165.93	0.0698	0.0459	0.0466	0.00258
(d)		P8	2144.80	1.34	0.129	0.0349	0.000357
(d)		P9	2123.24	0.187	0.0296	0.0169	0.00410
(d)		P10	2101.27	0.144	0.0322	0.0180	0.00599
(d)		P12	2056.14	0.114	0.0222	0.0131	0.000494
(d)		P13	2033.01	0.153	0.0198	0.00580	0.000100

Table 2.2 (continued)

	CO Laser Parameters			Atmospheric Absorption Coefficients (km^{-1})			
				z = 0 km, Sea Level			z = 12 km
	Band	Rot. ID	v(cm^{-1})	k_{trop}	k_{mw}	k_{sw}	k_{mw}
(d)	9-8	P6	2108.48	0.0603	0.0172	0.0119	0.00969
(d)		P7	2088.34	0.444	0.0567	0.0188	0.00663
(d)		P8	2067.76	0.791	0.112	0.0554	0.00259
(d)		P10	2025.36	0.646	0.0864	0.0253	0.000025

(a) Laser frequencies calculated using molecular constants of Young [51]
(b) Laser frequencies calculated using molecular constants of Mantz et al. [52]
(c) Measured, Deutsch [53]
(d) Calculated, Basov et al. [54]
(e) Measured, Spanbauer et al. [55]
(f) Calculated, Spanbauer et al. [55]

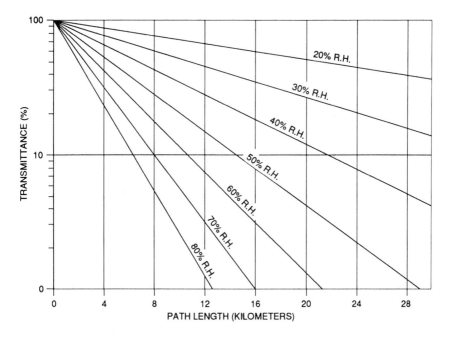

Figure 2.4 Calculated 10.59-μm transmittance of water vapor at 77°F in a sea-level atmosphere *versus* the path length for selected relative humidities [56].

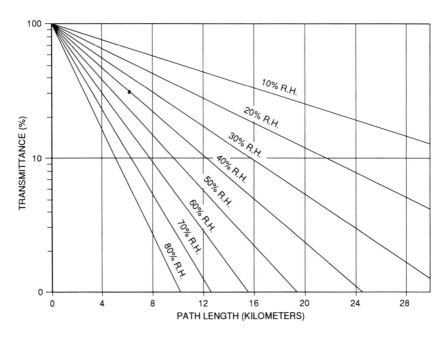

Figure 2.5 Calculated 10.59-μm (P20) transmittance of combined carbon dioxide and water vapor at 77°F in a sea-level atmosphere *versus* the path length [56].

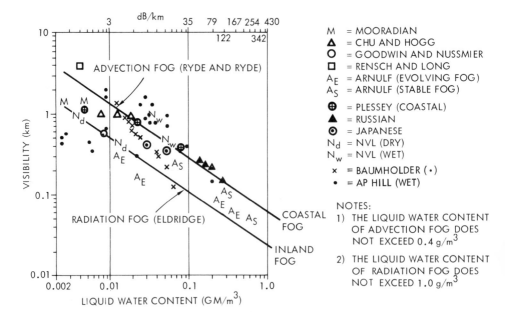

Figure 2.6 10-μm attenuation relative to microwave radar models (visibility $\leq$ 1 km) [57].

ADVECTION FOG: $\sigma_{10.6} = 1.08 \, v_1^{-1.64}$

RADIATION FOG: $\sigma_{10.6} = 0.24 \, v_1^{-1.64}$

Figure 2.7 10.6-μm extinction coefficient data fit (visibility ≤ 1 km) [57].

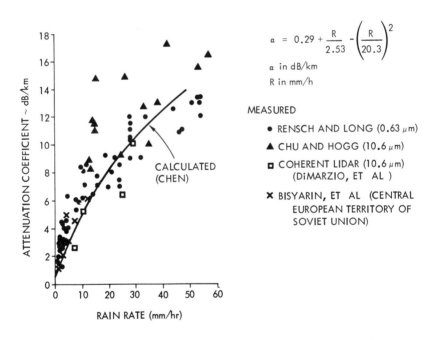

Figure 2.8 Measured and calculated attenuation coefficient *versus* rain date [58–61].

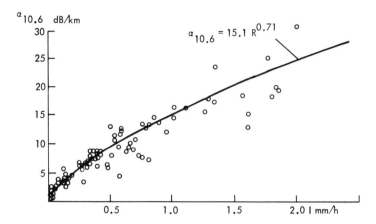

Figure 2.9 Attenuation in snow [62].

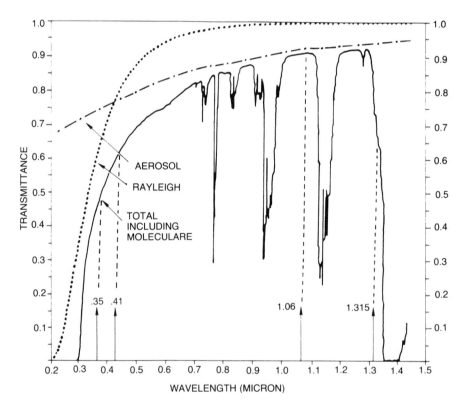

Figure 2.10 Propagation to space from 1.2 km, zenith midlatitude summer, 23-km visibility rural aerosol [33].

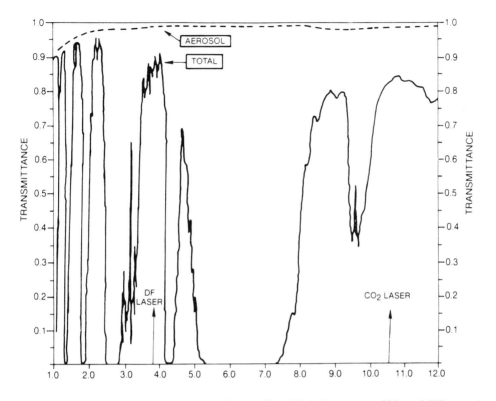

Figure 2.11 Propagation to space from 1.2 km, zenith midlatitude summer, 23-km visibility rural aerosol [33].

Table 2.3
Linear Propagation to Space [33]

	1.2-km altitude, vertical propagation, midlatitude summer atmosphere, 23-km visibility, rural aerosol	
Laser	*Wavelength (μm)*	*Transmission (%)*
XeCl	0.31	8
XeF	0.35	43
XeF + Raman	0.41	57
FEL	0.5–1.06	70–90
Nd:Yag	1.06	90
Iodine	1.315 (7603.1385)	87
HF $P_2(8)$	2.7 (3434.995)	42
DF (Avg)	3.8 (2500–2800)	86
CO_2P(20)	10.6 (944.1949)	44
$C_{13}O_2$	11.15 (896.9095)	86

Chapter 3
Detection Probabilities and False Alarm Rates

3.1 INTRODUCTION

In Chapter 1 the signal-to-noise relationship for incoherent and coherent detection systems was shown to be a function of a variety of noise sources. These noise sources can have an impact on the detection and false alarm rates of a system different than that of typical radar models. That some of the noise sources are non-Gaussian in nature, that targets may be smooth or diffuse, that the incoherent SNR is quadratically related to power, and that atmospheric turbulence can cause log-normal statistical fluctuation in the received signal intensity results in a number of special cases, are issues which must be considered.

Coherent detection systems, because of the strong local oscillator signal, are limited by local oscillator shot noise, resulting in detection statistics similar to those developed for microwave radars [49].

In order to appreciate these special conditions, the microwave radar models of Marcum and Swerling as portrayed by Skolnik will be reviewed. These models will then be shown to be the same for some coherent detection optical models in a nonturbulent atmosphere. The addition of turbulence and the consideration of target smoothness characteristics in the detection process, coupled with the nonnecessarily linear Gaussian noise receiver, will then result in a number of special conditions.

In order to illustrate the relationship between signal-to-noise ratios and detection probabilities, a summary of "The Mathematical Distinctions Between Microwave and Laser Photon Noise Limited Detection Statistics" by Drs. K. Seeber and G. Osche is included in Section 3.5, by permission [63]. In this analysis the authors show that receiver-target models, which assume classical Maxwellian nonquantum theory, are described by continuous probability distributions, consistent with the Correspondence Principal, where a physical system described by large quantum numbers is adequately treated by classical (nonquantum) theory. Applied to the radar/lidar case, this means that receivers having large numbers of photons can be described by microwave-radar statistical theory. In a heterodyne lidar the intense local oscillator field is superimposed on the received field, thus resulting in a large number of photons being detected even for weak signals. Microwave radar detection statistics (where the number of photons is always large because of the comparatively low frequency of the field) can therefore always be used for coherent lidar. For example, Swerling class targets used in radar detection theory can therefore be employed for laser systems.

The situation for direct detection lidar is not so clear cut: when the lidar receiver is sufficiently sensitive so that small numbers of photons can be detected, quantum statistics must be used, for systems that operate at high signal-to-noise levels continuous probabilities can be used. (In practice, large means more than 20 photons.) A further distinction between coherent and direct detection statistics is that coherent fields require two parameters to be specified (e.g., amplitive and phase) and the underlying statistics for the signal and the noise is two-dimensional (e.g., circulocomplex gaussian for the in-phase and quadrature components of the noise field). The direct detection process contains no phase information and is therefore completely characterized by a single parameter.

3.2 TARGET DETECTION STATISTICS (MICROWAVE RADAR MODELS)

The detection of a target having a fluctuating signal intensity in a Gaussian noise receiver of a microwave search radar system (i.e., point target having an inverse fourth power of range relationship) was analyzed by Marcum and Swerling [64,65]. In these Rand Corporation reports, P. Swerling identified four different models of fluctuating targets and one model of a nonfluctuating target. The following material is extracted from this report in order to allow the reader to understand the models as described in Cases 1 to 4 below.

> Four different models of target fluctuation were considered by Swerling [65]. The four models chosen for consideration were felt to be typical of situations which are likely to be met in practice, or, at least, to bracket a wide range of practical cases.
>
> In applying probability of detection computations to actual cases, one should first attempt to analyze the fluctuations of the actual target under consideration, and then choose whichever model (including Marcum's non-fluctuating model) appears to most closely approximate the actual target fluctuations. Or, one could consider the actual target to be intermediate between two of the theoretical models.
>
> One of the uses of the ensuing results is, that by comparing the results for different models, one can make some judgement as to the errors introduced by choosing the wrong fluctuation model.
>
> The four fluctuation models considered are as follows:

Case 1

The returned signal power per pulse is assumed to be constant for the time on target during a single scan, but to fluctuate independently from scan to scan. (This ignores factors such as beam shape effect.) Expressed in statistical terms, the normalized autocorrelation function of target cross section is approximately one for the time in which the beam is on target during a single scan, and is approximately zero for a time as long as the interval between scans. This type of fluctuation will henceforth be referred to as scan-to-scan fluctuation.

The fluctuations of target cross section are evidenced as fluctuations of signal-to-noise ratio in the receiver. The probability density for the input signal-to-noise power ratio is assumed to be

$$w(x,\bar{x}) = \frac{1}{\bar{x}} e^{\frac{x}{\bar{x}}} \; (x \geq 0)* \tag{3.1}$$

where

x = input signal-to-noise power ratio
$\bar{x}$ = average of x over all target fluctuations

Case 2

The fluctuations are independent from pulse to pulse. This type of fluctuation will be referred to as pulse-to-pulse fluctuation.

The probability density function is given by Equation (3.1).

Case 3

Scan-to-scan fluctuation according to the probability density:

$$w(x,\bar{x}) = \frac{4x}{\bar{x}^2} e^{-2x/\bar{x}} \; (x \geq 0) \tag{3.2}$$

Case 4

Pulse-to-pulse fluctuation according to Equation (3.2).

It would be well at this point to indicate in which actual situations the various models would be likely to apply.

---

*This formula also represents the probability density for target cross section Σ if x is replaced by Σ and $\bar{x}$ by $\overline{\Sigma}$.

As to the choice of probability density function for the fluctuations:

Theoretically, for a target which can be represented as several independently fluctuating reflectors of approximately equal echoing area, the density function should be close to exponential, even if the number of reflectors is as small as four or five. Thus one would expect objects which are large compared with wavelength (and which are not shaped too much like a sphere) to fluctuate approximately according to the exponential density of Equation (3.1).

On the other hand, a target which can be represented as one large reflector together with other small reflectors, or as one large reflector subject to fairly small changes in orientation, would be expected to behave more like Equation (3.2).

The non-fluctuating model would apply to spherical or nearly spherical objects (e.g., balloons, meteors) at fairly large wavelengths.

Most available microwave observational data on aircraft targets indicates agreement with the exponential density, Equation (3.1). More definite statements as to actual targets for which Equation (3.2) of the non-fluctuating model apply must await further experimental data.

As to the choice between scan-to-scan and pulse-to-pulse fluctuation:

The scan-to-scan model would apply to targets such as jet aircraft or missiles, for radars having fairly high pulse repetition rate and scan rate.

Pulse-to-pulse fluctuation would apply to propeller-driven craft if the propellers contribute a large portion of the echoing area, or to targets for which very small changes in orientation would mean large changes in echoing area, such as long, thin objects at high frequency; or to targets viewed by a radar with sufficiently low repetition rate.

Most actual targets would probably be intermediate between the various cases considered.

A comparison between Cases 1, 2, 3, 4, and the non-fluctuating case is given in Figure 3.1 (for typical false alarm time and number of hits integrated).

In all cases, it is assumed that there are on each scan N hits; after passage through a square law second detector, the resulting N pulses are added and required to exceed a threshold Y_b in order for detection of a target to occur.* The second detector output is, for mathematical convenience, assumed to be normalized as follows: detector output equals

---

*The actual beam shape is in effect being approximated by a beam having uniform gain over a finite sector, and zero gain outside this sector. In principle, it is possible to take exact account of beam shape in computing probability of detection. This is not done here, however, since it is through that that the effect of the aforementioned approximation is small, provided the effective number N of hits per scan and effective $\bar{x}$ are properly chosen.

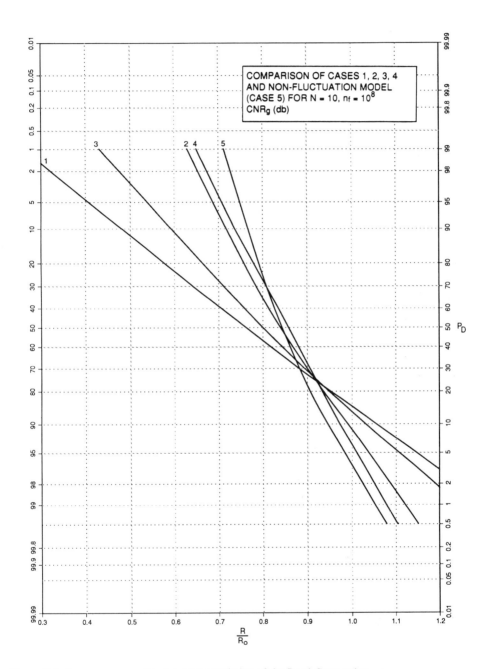

Figure 3.1 P_D *versus* range. Reprinted by permission of the Rand Corporation.

input envelope squared divided by twice the mean square input noise voltage.**

M. Skolnik in *Introduction to Radar Systems* [49] converted the probability of detection figures of Swerling's paper (Fig. 3.1) from a P_D versus R/R_o (normalized

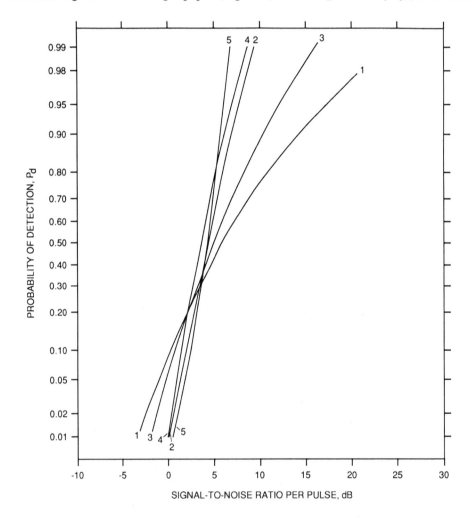

Figure 3.2 Comparison of detection probabilities for five different models of target fluctuation for $N = 10$ pulses integrated and false-alarm number $n_f = 10^8$ ($n_f = n/P_{fa}$). (Adapted from Swerling, reprinted by permission of McGraw-Hill [49].)

**Author's Note: At the time of this publication by Marcum, the microwave radar data base was still being compiled. Since that time, substantial experimental data has been collected.

range) curve to P_D versus SNR by scaling the curves with the relationship that the SNR varied as $1/R^4$, which is observed in Figure 3.2. This allowed a more general utilization of the data. For the microwave radar case (Figure 3.2) where 10 pulses are integrated and a false alarm number of 10^8 is selected, the SNR versus P_D effects of the various models may be clearly visualized. Figures 3.3 and 3.4 from Skolnik illustrate the additional SNR requirements and integration improvement factors for probability of detection for the five Swerling targets.

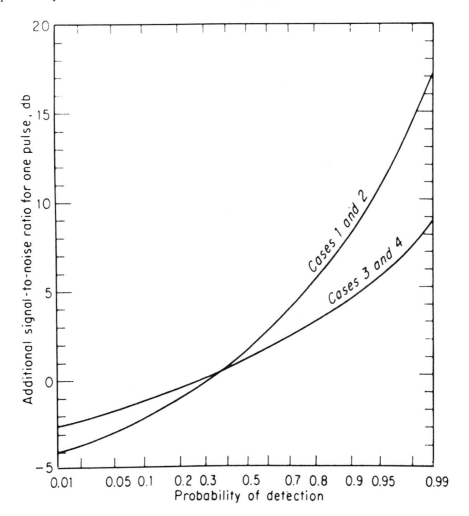

Figure 3.3 Additional SNR required to achieve a particular probability of detection, when the target cross section fluctuates, as compared with a nonfluctuating target; single hit, $N = 1$. (Reprinted by permission of McGraw-Hill [49].)

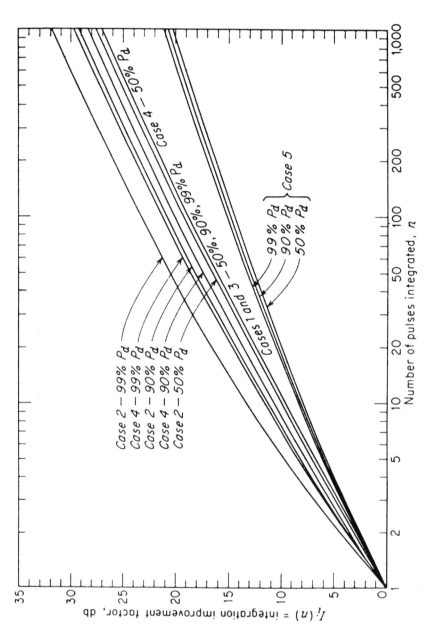

Figure 3.4 Integration improvement factor as a function of the number of pulses integrated for the five cases of target fluctuations considered. (Reprinted by permission of McGraw-Hill [49].)

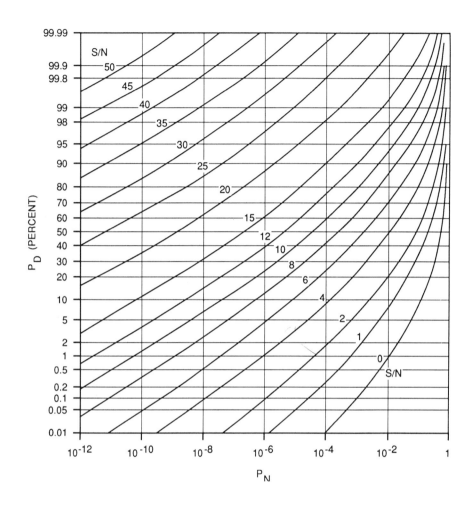

Figure 3.5 P_D versus P_{FA} (sinusoidal). (Reprinted by permission of Raytheon Co. [67].)

Figure 3.5 illustrates the probability of detection (P_D) *versus* probability of false alarm (P_{FA}) for a single look at a sinusoidal (nonfluctuating) signal in noise, and at a Gaussian (Figures 3.6 and 3.7) signal in noise where S is the mean signal power, N is the mean noise power, and P_D equals $P_N^{1/(1+S/N)}$ [67].

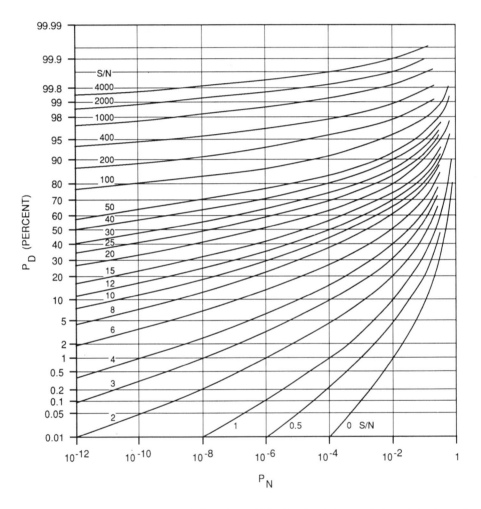

Figure 3.6 P_D versus P_{FA} (Gaussian) SNR = 0 to 4000. (Reprinted by permission of Raytheon Co. [67].)

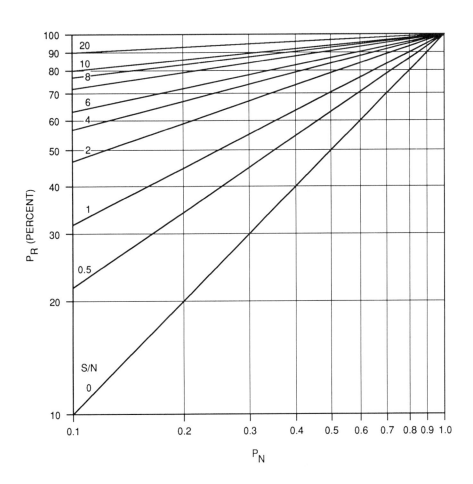

Figure 3.7 P_D versus P_{FA} (Gaussian) SNR = 0 to 20. (Reprinted by permission of Raytheon Co. [67].)

3.3 COHERENT DETECTION LASER RADAR MODELS

3.3.1 Introduction

The microwave radar detection models developed by Marcum and Swerling have equal applicability at optical wavelengths. However, care must be exercised in the arbitrary use of these models because the optical system has different receiver noise mechanizations and is more substantially affected by the atmospheric structure, target surface characteristics, decorrelation time, and the spatial and temporal coherence of the laser system.

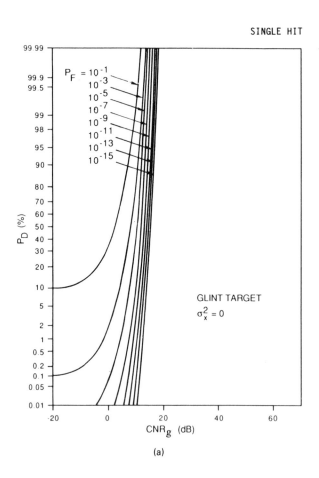

(a)

Figure 3.8 SNR requirements—probability of detection and false alarm (a) glint/(b) speckle targets [68].

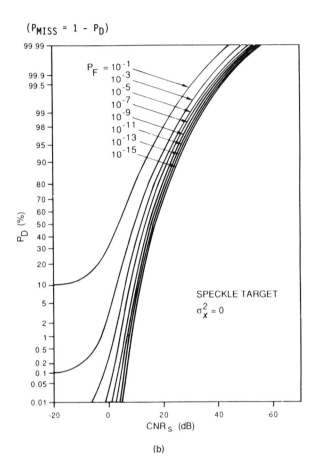

Figure 3.8 (cont'd)

Figure 3.8 illustrates the probability of detection and false alarm curves computed for an optically coherent receiver where the targets are classified as "glint" (nonfluctuating) and "speckle" (diffuse) type. In the absence of atmospheric turbulence, these models are representative of a sinusoidal-Swerling I target (in Gaussian noise) and a Swerling II (Gaussian signal) microwave radar model, respectively. This may be seen by comparing Figures 3.8 and 3.9 with Figures 3.5 and 3.6. In the referenced work, the authors used carrier-to-noise ratio (CNR) as the direct equivalent to the microwave radar definition of SNR where CNR_g and CNR_s refer to glint (g) and speckle (s) target models. In order not to create confusion in translation, both notations will be used as originated by the respective authors.

DETECTION PROBABILITY	FALSE ALARM PROBABILITY - P_F							
P_D	10^{-1}	10^{-3}	10^{-5}	10^{-7}	10^{-9}	10^{-11}	10^{-13}	10^{-15}
.0001	**	**	-6.02	-1.25	0.97	2.43	3.52	4.39
.0010	**	$-\infty$	-1.76	1.25	3.01	4.26	5.23	6.02
.0100	**	-3.01	1.76	3.98	5.44	6.53	7.40	8.13
.1000	$-\infty$	3.01	6.02	7.78	9.03	10.00	10.79	11.46
.3000	-0.40	6.76	9.33	10.93	12.10	13.02	13.78	14.42
.5000	3.66	9.53	11.93	13.47	14.61	15.51	16.25	16.89
.7000	7.37	12.64	14.95	16.45	17.57	18.45	19.19	19.82
.8000	9.69	14.76	17.04	18.53	19.63	20.51	21.24	21.87
.8500	11.20	16.18	18.44	19.92	21.02	21.90	22.63	23.25
.9000	13.19	18.10	20.35	21.82	22.92	23.79	24.52	25.14
.9500	16.42	21.26	23.49	24.96	26.05	26.93	27.65	28.28
.9800	20.53	25.33	27.55	29.01	30.11	30.98	31.70	32.33
.9900	23.58	28.37	30.59	32.05	33.14	34.01	34.74	35.36
.9950	26.61	31.39	33.61	35.07	36.16	37.03	37.76	38.38
.9990	33.62	38.39	40.61	42.07	43.16	44.03	44.76	45.38
.9999	43.62	48.39	50.61	52.07	53.16	54.04	54.76	55.38

CNR_s in dB

20 dB

Capron, Harney, Shapiro, Report TST-33 MIT/LL, July 1979.

Figure 3.9 SNR requirements—target receiver operating characteristics $\sigma_x^2 = 0$ [68].

3.3.2 Atmospheric Turbulence Effects on Laser Radar Model

When optical energy is propagated through the atmosphere [69], it is affected by the refraction index of the atmosphere, which depends upon pressure, temperature, humidity, and wavelength. The largest impact is that caused by temperature effects. Turbulence in the atmosphere results in a nonhomogeneous refraction index over a path and results in spatial and temporal variations in the propagation beam. In the time domain, turbulence will result in variations of the received signal strength and spatial effects. Spatial effects will result in beam distortion, beam wander, spatial coherence limitations, focusing, and spreading of the optical beam (a complete description of these effects is included in an article by Fante [70]).

The atmospheric fluctuations [2] cause the optical beam to wander from its projected path and induces scintillation in the received signal. The variances of these fluctuations [σ_x^2] and [σ_S^2], respectively are dependent upon the atmospheric phase structure function (C_N), the path length (R), altitude z aperture radius ($D/2$), where

$$\sigma_S^2 = 3.44 \left(\frac{D}{2}\right)^{-7/3} \int_O^R (R-z)^2 C_N^2 \, dz \quad (3.3)$$

For $D/2$ = large and

$$\sigma_X^2 = 0.31 \, C_N^2 \, K_v^{7/6} \, R^{11/6} \text{ Plane wave} \quad (3.4)$$

$$\sigma_X^2 = 0.124 \, C_N^2 \, K_v^{7/6} \, R^{11/6} \text{ Spherical wave} \quad (3.5)$$

where K_v is the wave vector $\equiv 2\pi/\lambda$ (m^{-1}).

The effects of beam wander and scintillation on a coherent scanning laser radar system, where the intensity of the received signal from a scene composed of NXN pixels was used to create an image, was evaluated by Paport, Shapiro, and Harney [71]. In this work, the probability of detection and false alarm curves were generated (Figures 3.10 to 3.13) for both glint and diffuse (speckle) targets when subjected to various levels of scintillation and aperture averaging of the beam wander and to the number of pulses integrated (I). For a 20-dB CNR, one can observe a substantial reduction in the detection probability from the no-turbulence condition. Similar effects are also observed for the probability of detection statistics as a function of SNR for speckle, diffuse, and semirough targets, which are illustrated in Figure 3.14 [72B].

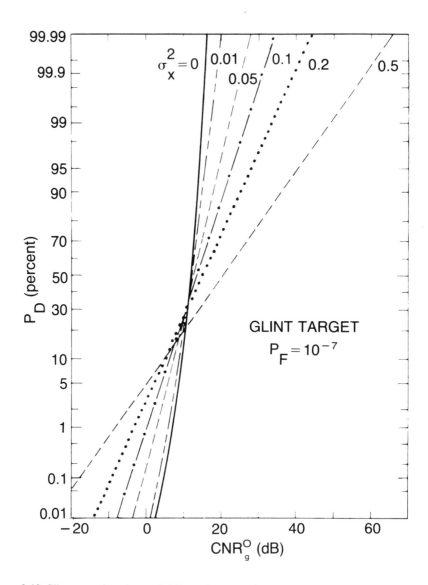

Figure 3.10 Glint target detection probability. (Courtesy of SPIE.)

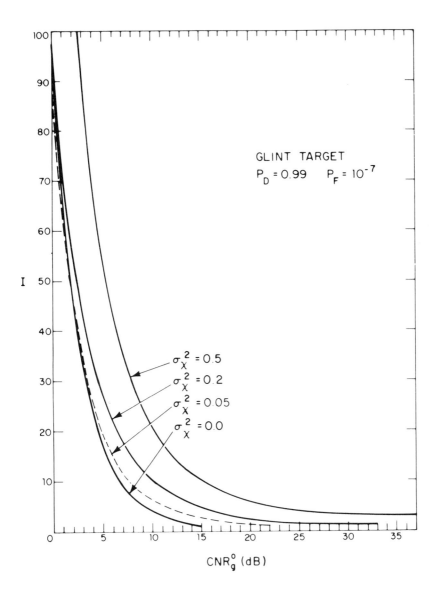

Figure 3.11 Glint target pulse integration. (Courtesy of SPIE.)

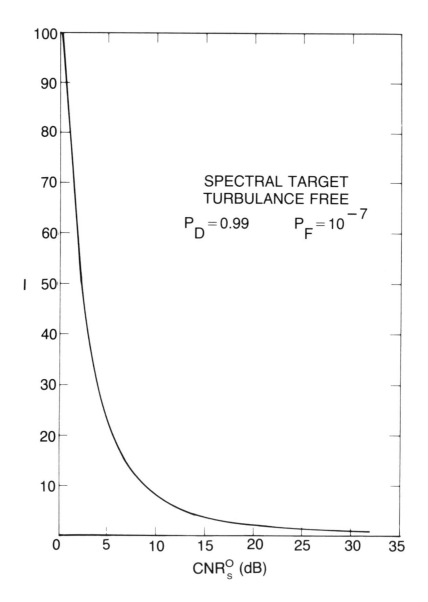

Figure 3.12 Speckle target pulse integration. (Courtesy of SPIE.)

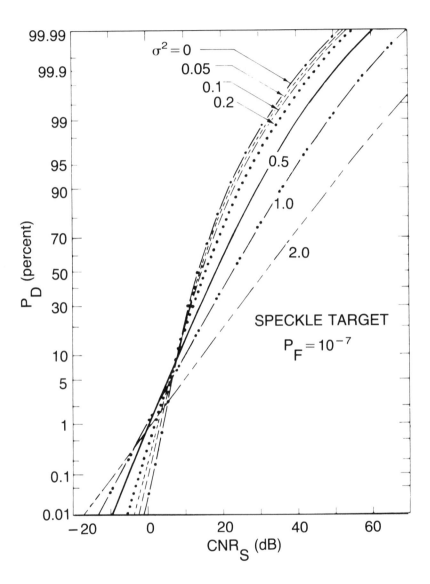

Figure 3.13 Speckle target detection probability. (Courtesy of SPIE.)

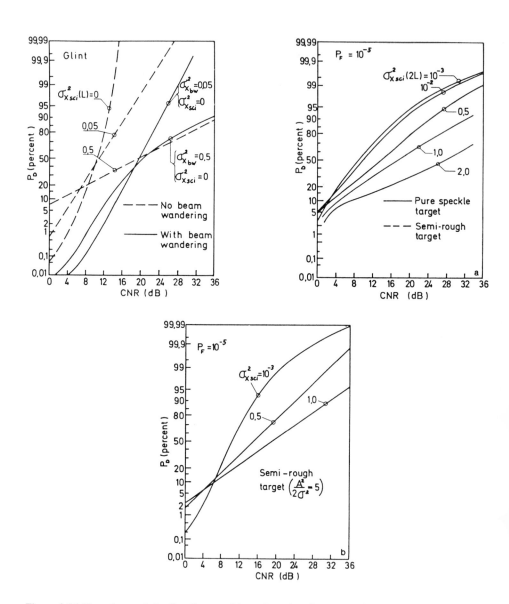

Figure 3.14 Detection statistics for glint, speckle, and semirough targets. (Courtesy of *Applied Optics*.) D. Letalick, I. Renhorn, O. Steinvall, "Target and Atmospheric Influence on Coherent CO_2 Laser Radar Performance," *Applied Optics*, Vol. 25, No. 21, Nov. 1986, pp. 3939–3945.

3.4 PROBABILITY OF DETECTION AND FALSE ALARM RATES FOR INCOHERENT DETECTION SYSTEMS

3.4.1 Introduction

At low photon or photoelectron levels, the detection of a non-fluctuating signal in background noise is determined by Poisson statistics. This condition applies to the energy detection or incoherent detection receiver. Here, the system designer must compute the average number of the noise photoelectrons emitted ($\bar{n}_N$) in the time interval of interest, as well as the average number of signal photoelectrons in the time interval of interest ($\bar{n}_s$), in order to set the decision threshold detecting device at a level of photoelectrons (n_T) corresponding to a distinct probability of detection (P_D) and probability of false alarms (P_{FA}). For the Poisson statistics case, the probability of detection may be expressed as [16,42,72A]

$$P_D = \sum_{r=R_t}^{\infty} P(S+N)\tau = \sum_{r=R_t}^{\infty} \frac{(n_s + n_N)^r}{r!} \exp^{-(\bar{n}_s + \bar{n}_N)} \qquad (3.6)$$

where $P(S+N)$ = the probability that r signal plus noise photoelectrons are emitted in time interval, τ, and the probability of false alarm may be expressed as

$$P_{FA} = \sum_{r=R_t}^{\infty} \frac{(\bar{n}_N)^r}{r!} \exp^{(-\bar{R}_N)} \qquad (3.7)$$

Figures 3.15 and 3.16 permit the computation of P_D and P_{FA}, respectively. The following example illustrates the use of these curves.

Laser Detection Statistics

Given: Signal photoelectrons = 6
Noise photoelectrons in the absence of signal = 1
Pulse width, τ, = 20 ns (video detection)
Average false alarm rate = 1 in 10^3
Problem: What is probability of detection?

For Figure 3.16, enter $P_{FA} = 10^{-3}$, with $\bar{n} = 1$ photoelectron; obtain threshold setting $n_\tau = 5.8$ photoelectrons ≈ 6 photoelectrons.
For Figure 3.15, enter $n_\tau = 5.8$ photoelectrons and $\bar{n} = 7$ photoelectrons (6 signal + 1 noise); obtain $P_D \approx 0.7$.

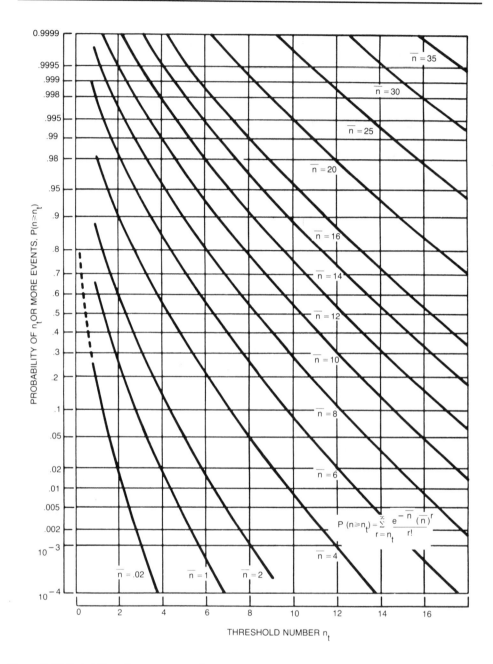

Figure 3.15 Probability of n_r or more random events with Poisson distribution when the expected (or mean) number of events is $\bar{n}$ as a function of the threshold number n_r. (From *RCA Electro-Optics Handbook* [4].)

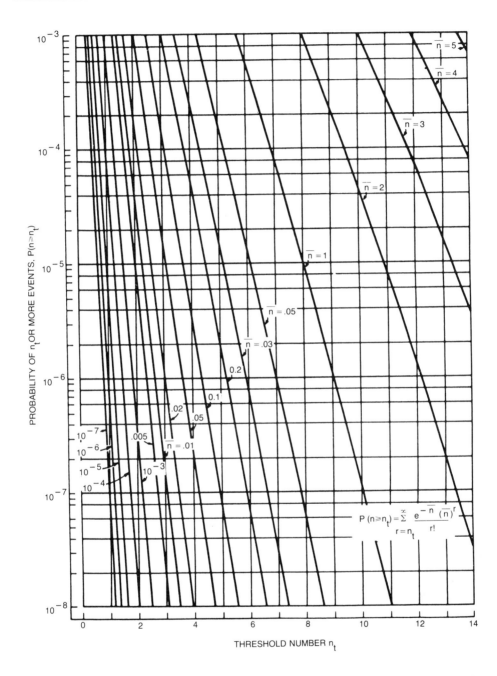

Figure 3.16 Probability of n_τ or more random events with Poisson distribution when the expected (or mean) number of events is $\bar{n}$ as a function of the threshold number n_τ (curves for $10^{-7} \leq n \leq 10^{-4}$ are approximate). (From *RCA Electro-Optics Handbook* [4].)

3.4.2 Incoherent Detection of Optically Noise (Poisson) Limited Receivers

Goodman [72,73] considered the impact of target smoothness characteristics upon the detection statistics for an incoherent detection receiver collecting coherent radiation from both specular and diffuse (speckle) targets. Figure 3.17 illustrates the energy detection radar modeled.

In this analysis, the receiving aperture was considered as having a number (M) of independent spatial *correlation cells* (speckle), where the energy could be considered constant within a cell and statistically independent of the energy density in all other "correlation cells." Specular target detection statistics were shown to be represented by Poisson statistics, while "rough" target signal statistics were represented by negative binomial statistics. For $M = 1$ (i.e., a single speckle), these statistics reduced to Bose-Einstein. When the average number of signal photoelectrons ($\bar{N}_s$) per spatial correlation cell was much less than one, the detection statistics could be modeled as Poisson for the rough target.

Figures 3.18 and 3.19 illustrate, respectively, a specular and rough target probability P_D versus N_s for a family of average number of noise photoelectrons ($\bar{N}_N$) for a probability of false alarm of 10^{-6}.

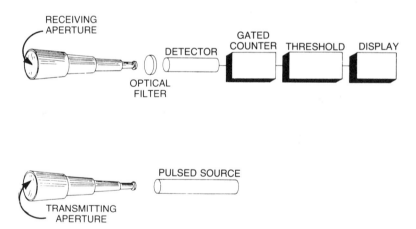

Figure 3.17 Energy-detection radar, © IEEE. (Courtesy of IEEE.)

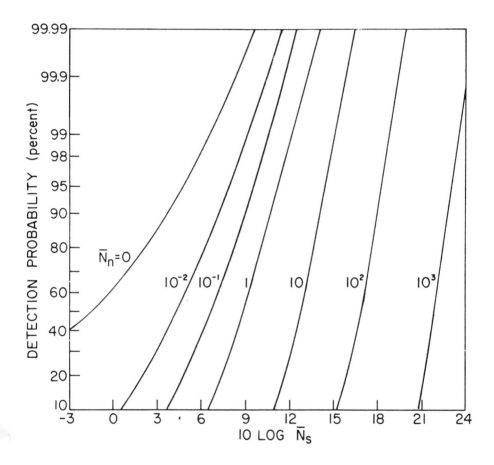

Figure 3.18 Energy-detection radar performance—specular target $P_{FA} = 10^{-6}$. (Courtesy of IEEE.)

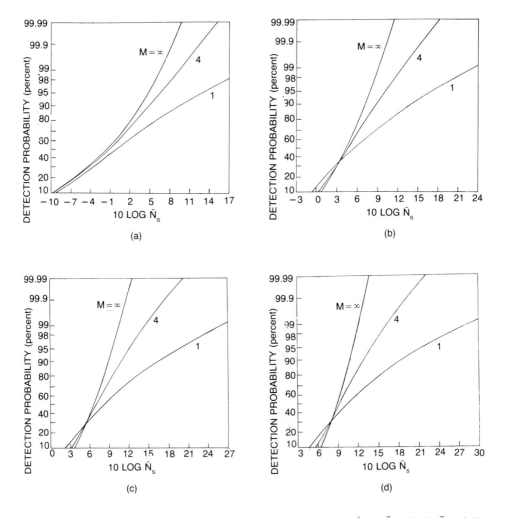

Figure 3.19 Energy detection radar performance—rough target $P_{FA} = 10^{-6}$, (a) $\bar{N}_n = 0$, (b) $\bar{N}_n = 0.01$, (c) $\bar{N}_n = 0.1$, (d) $\bar{N}_n = 1.0$, (e) $\bar{N}_n = 10.0$; © IEEE. (Courtesy of IEEE.)

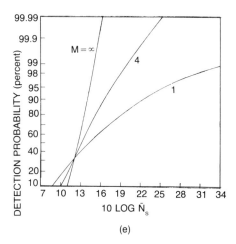

(e)

Figure 3.19 (cont'd)

3.4.3 Incoherent Detection of Thermal (Gaussian) Noise-Limited Receivers

Seeber and DiMarzio [74] in their unpublished work titled "Incoherent Optical Signal Detection in Gaussian Noise" evaluated the detection of speckle targets when photon noise was not the dominant receiver noise term. In this work they point out that—

- Aperture averaging does not average the dominant noise term.
- Detection probabilities are calculated from the optical power, received, which, in the incoherent case, is proportional to the square root of the SNR. This dependance on the optical power is the same as in the case of coherent detection, but the probability functions are different.

Figures 3.20, 3.21, and 3.22 from their work illustrate the probability of detection and false alarm curves for 1, 10, and 100 aperture-averaged speckles, respectively, while Figure 3.23 illustrates the improvement associated with sample averaging 10 aperture-averaged speckles. This may be observed in Table 3.1.

Figure 3.24 illustrates the coherent and incoherent detection signal noise required to obtain the same detection probability for a false alarm of 10^{-6}. It can be observed that if a 20-dB SNR were required for the coherent detection statistics, equivalent statistics would necessitate an SNR of approximately 35 dB.

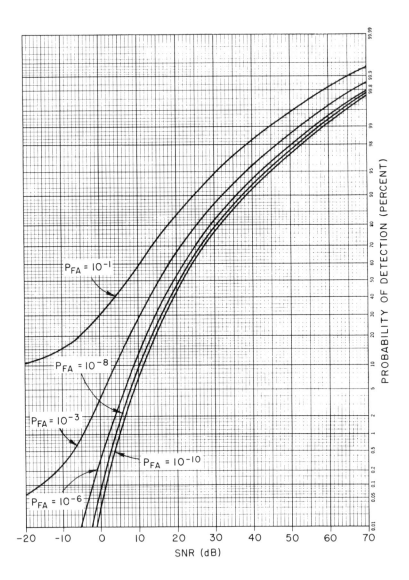

Figure 3.20 Single speckle diffuse target for Incoherent Gaussian Noise Receiver (IGNR). (Courtesy of Raytheon Co.)

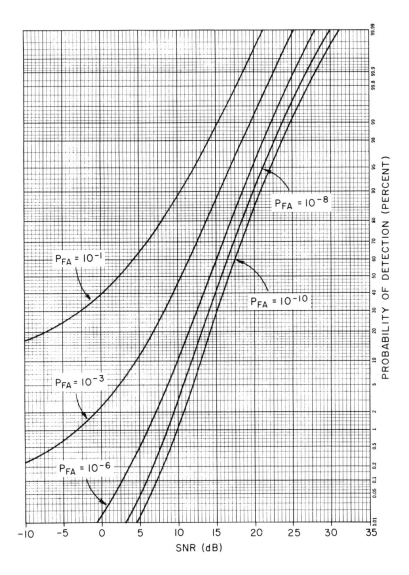

Figure 3.21 10 speckles aperture averaged for (IGNR).

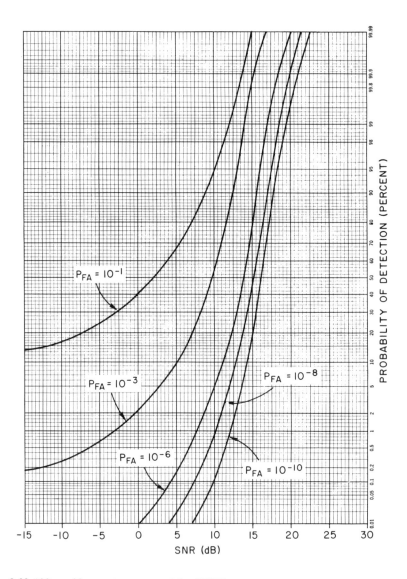

Figure 3.22 100 speckles aperture averaged for (IGNR).

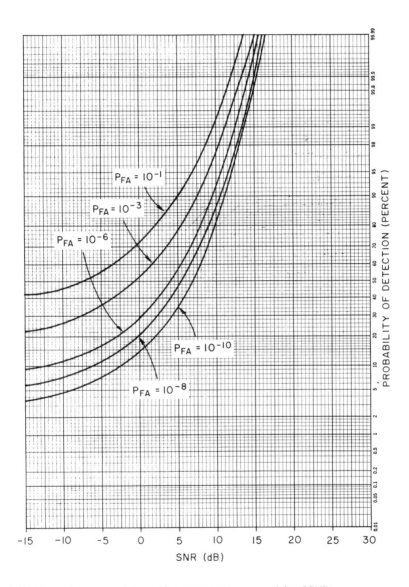

Figure 3.23 10 samples averaged times 10 measurements averaged for (IGNR).

Table 3.1
Detection Probabilities for Incoherent Gaussian Noise Receiver

SNR (dB)	No. of Speckles Averaged	P_{FA}	P_D
20	1	10^{-10}	0.45
20	10	10^{-10}	0.85
20	100	10^{-10}	0.99.9
20	10 Samples of 10	10^{-10}	>0.99.99

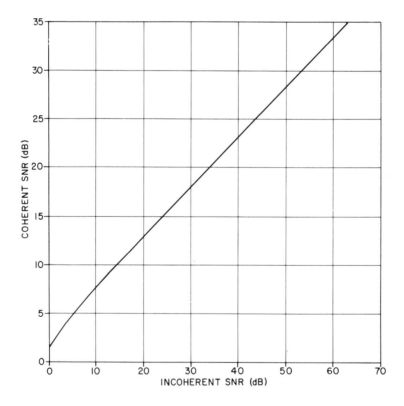

Figure 3.24 Comparison between coherent and incoherent SNR required for identical P_D for $P_{FA} = 10^{-6}$ single-sample SNR for (IGNR).

3.5 MATHEMATICAL DISTINCTIONS BETWEEN MICROWAVE AND LASER PHOTON NOISE LIMITED DETECTION STATISTICS

3.5.1 Introduction

The purpose of this section is to summarize the mathematical differences between Swerling II and Swerling IV, Rician, and log normal statistics. It should be noted that Swerling I and III use the same probability density (PD) function for a single sample as Swerling II and IV, respectively. They differ only for more than one sample.

3.5.2 Swerling II

The probability density function for the amplitude of a Swerling II target is given by

$$p(a) = \frac{2a}{a_o^2} \exp\left(-\frac{a^2}{a_o^2}\right) \quad \text{for} \quad a \geq 0 \tag{3.8}$$

$$p(a) = 0 \quad \text{for} \quad a < 0$$

where a_o is the average amplitude $\langle a \rangle$, and a is the amplitude.

When the square of the amplitude (I) is used as the statistical variable, the PD is transformed into

$$p(I_1) = \frac{I_1}{I_o} \exp\left(-\frac{I_1}{I_o}\right) \quad \text{for} \quad I_1 \geq 0 \tag{3.9}$$

$$p(I_1) = 0 \quad \text{for} \quad I_1 < 0$$

where

I_1 = the amplitude squared
I_o = the average of the amplitude squared
$p(I_1)$ = the probability density of the amplitude squared

$$I_o = a_o^2 = \langle 1 \rangle \tag{3.10}$$

The PD of random noise is also given by Equation (3.8) or (3.9), and the detection probability for Swerling II signal in random noise is given by

$$P_D = P_{FA}^{\frac{1}{1+SNR}} \tag{3.11}$$

where P_{FA} is the false alarm rate and SNR is the signal-to-noise ratio.

Frequently, the signal power is plotted on a logarithmic scale. By putting $Z = \ln I_1/I_o$, Equation (3.9) is transformed into

$$p(Z) = 2 \exp(Z) \exp(-\exp(2Z)) \tag{3.12}$$

If a different scale is used:

$$Z' = \ln K \frac{I_1}{I_o} \tag{3.13}$$

the origin of the Z axis is simply shifted by $\ln K$.

3.5.3 Swerling IV

The PD for a Swerling IV amplitude is given by

$$p(a) = \frac{8a^3}{a_o^4} \exp\left(-\frac{2a^2}{a_o^2}\right) \tag{3.14}$$

and for the power by

$$p(I_1) = \frac{4I_1}{I_o^2} \exp\left(-\frac{2I_1}{I_o}\right) \tag{3.15}$$

The log distribution of this power is

$$p(Z) = 8 \exp(3Z) \exp(-2 \exp(2Z)) \tag{3.16}$$

with

$$Z = \ln \frac{I_1}{I_o} \tag{3.17}$$

The probability of detection for a Swerling IV signal in random noise $p(y)$ is given by

$$P_D = \int_O^T p(y)dy \qquad (3.18)$$

where y is the signal intensity plus noise intensity and where

$$p(y) = \left(1 + \frac{S}{2N}\right)^{-2} \exp(-y){}_1F_1\left(2, 1, \frac{1}{1 + \frac{S}{2N}}\right) \qquad (3.19)$$

where ${}_1F_1$ is the confluent hypergeometric function. T_V is the threshold and is related to the false alarm rate by

$$T_V = \ln P_{FA} \qquad (3.20)$$

3.5.4 Comparison of Swerling II and IV

The distribution functions for all cases of interest are plotted in Figures 3.25 through 3.30. Figures 3.25 and 3.26 show the amplitude distribution for Swerling II and IV, respectively. Figures 3.29 and 3.30 show the power distribution functions. (All values are normalized with $a_o = 1$ and $I_o = 1$.) The logarithmic distributions are shown in Figures 3.27 and 3.28. It is readily apparent that Swerling IV targets have a much

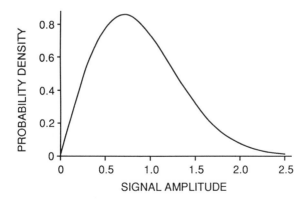

Figure 3.25 Swerling I and II amplitude distribution.

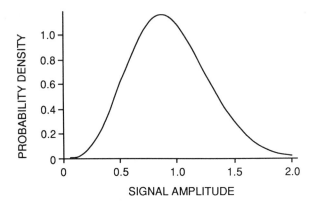

Figure 3.26 Swerling III and IV amplitude distribution.

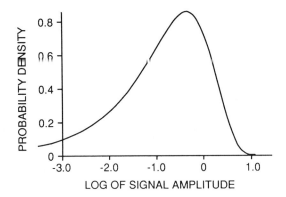

Figure 3.27 Swerling I and II log amplitude distribution.

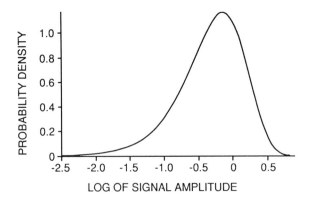

Figure 3.28 Swerling III and IV log amplitude distribution.

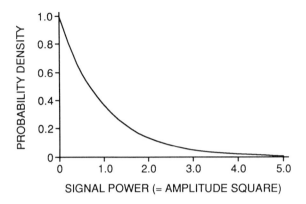

Figure 3.29 Swerling I and II power distribution.

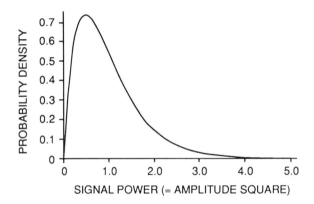

Figure 3.30 Swerling III and IV power distribution.

Table 3.2
Comparison of Swerling II and Swerling IV Targets

P_d	P_{FA}	SNR (Swerling II)	SNR (Swerling IV)
0.90	10^{-8}	22.4	10.4
0.95	10^{-8}	25.5	12.2
0.99	10^{-8}	32.7	16.5

smaller probability of fading (small signals) than Swerling II targets. This implies that a smaller SNR is required for Swerling IV in order to obtain a desired detection probability when a false alarm rate is given. These results are born out by the values in Table 3.2.

3.5.5 The Log-Normal Distribution

The log-normal density function is given by

$$p(a) = \frac{1}{\sqrt{2\pi}\sigma_L a} \exp\left[-\frac{(\ln a - \mu)^2}{2\sigma_L^2}\right] \qquad (3.21)$$

with $\mu = \ln \sigma_m$, where σ_L is the standard deviation of $\ln a/\sigma_m$ and σ_m the median of a.

The ratio of the mean-to-median value of a, ρ_l is related to the standard deviation by

$$\sigma_L = (2 \ln \rho_l)^{1/2} \qquad (3.22)$$

(where σ_L is the log of the amplitude) which is a measure of the width of the log-normal distribution. A plot of Equation (3.21) is shown in Figures 3.31 and 3.32.

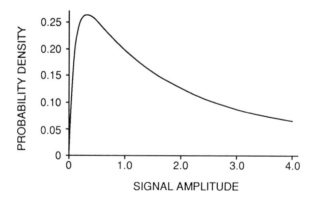

Figure 3.31 Log-normal distribution function.

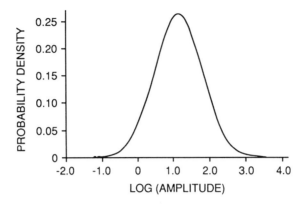

Figure 3.32 Log-normal amplitude distribution.

3.5.6 The Rice Distribution

The Rice distribution represents the statistical behavior of two target signals of interest: a glint in receiver noise and a glint plus diffuse target signal in noise. The Rician density is given by

$$p(v) = v_S \exp\left[-\frac{v_S^2 + a^2}{2}\right] I_o(av) \quad (3.23)$$

where

$$v_S = \frac{\text{amplitude}}{(\text{noise power} + \text{diffuse signal power})^{1/2}} \quad (3.24)$$

$$a = \left[\frac{2 \text{ glint power}}{\text{noise power} + \text{diffuse signal power}}\right]^{1/2} \quad (3.25)$$

To compute detection probabilities, the density function must be integrated between a threshold T and infinity. In terms of the variable v, the lower integration limit is given by

$$V_t = \left(\frac{2 P_{TV}}{N + S_a}\right)^{1/2} \quad (3.26)$$

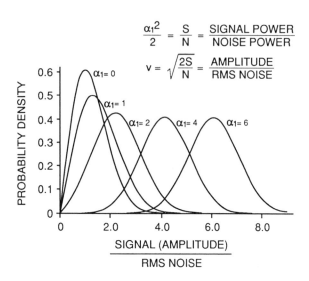

Figure 3.33 Rician distribution function.

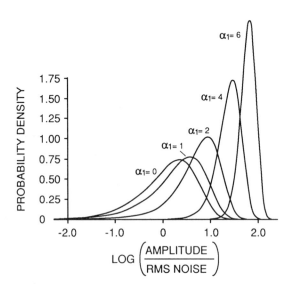

Figure 3.34 Rician log amplitude distribution.

where

$P_{TV} = \ln P_{FA}$ (P_{FA} = the probability of false alarm)
N = noise power
S_a = diffuse signal power

The Rician distributions for various values of α_1 are shown in Figures 3.33 and 3.34.

3.6 LIDAR SIGNAL STATISTICS IN THE PHOTON COUNT LIMIT

When the signal energy received corresponds to a small number of photons, classical statistics as used in the previous section are no longer valid. The correct probability distributions are derived from quantum mechanics. Each classical distribution has its quantum mechanical counterpart. A brief summary of these probability distributions is given in Table 3.3.*

Table 3.3
Radar and Lidar Signal Statistics

Radar	Lidar		Statistics
Constant Amplitude and Phase	Glint:	Coherent state	Poisson
Rayleigh Amplitude; Uniformly Distributed Phase	Speckle:	Chaotically Excited single mode	Bose-Einstein (geom. distribution)
Rayleigh Fluctuating Signal Plus Constant Signal; Rician Statistics	Speckle and Glint:	Coherent state plus noise-excited single mode	Laguerre
Incoherently Integrated Rayleigh Signal (Gamma Distribution)	Wideband chaotic field		Negative binomial

The following is a concise mathematical description of the various functions.

1. *Deterministic Field: Poisson Distribution*

   We define the average counting rate (W_p) (e.g., photodetections per second) then the average number of photo events ($\bar{K}$) in the time t, is given by

   $$\bar{K} = W_p t \qquad (3.27)$$

---

*D. Youmans has pointed out that Teich and Salen [104] have shown that these distributions apply to photoelectrons as well as photons.

The probability of counting n events is then given by

$$p(n,t) = \frac{\bar{K}^n}{n!} \exp(-\bar{K}) \qquad (3.28)$$

2. *Narrowband Optical Gaussian Field: Bose Einstein Distribution*
This corresponds to single speckle detection. The photon counting distribution for a speckle target echo is given by

$$p(n,t) = \frac{(\bar{K})^n}{(1+n)^{n+1}} \qquad (3.29)$$

The variance is given by

$$\sigma_K^2 = \bar{K} + (\bar{K})^2 \qquad (3.30)$$

3. *Wideband Optical Gaussian Field: Negative Binomial Distribution*
The photon count distribution for M uncorrelated speckles is given by

$$p(n,t) = \frac{\Gamma(n+M)}{\Gamma(n+1)\,\Gamma(M)} \left[1 + \frac{M}{\bar{K}}\right]^{-n} \left[1 + \frac{\bar{K}}{M}\right]^{-M} \qquad (3.31)$$

where Γ denotes the gamma function. The variance is equal to

$$\sigma_K^2 = \bar{K}\left(1 + \frac{\bar{K}}{M}\right) \qquad (3.32)$$

4. *Deterministic Field Plus Narrowband Gaussian Field:* This case corresponds to the return signal being composed of a glint and diffuse target return. Here the counting rate is now the sum of two terms:

$$W_P = W_G + W_S \qquad (3.33)$$

and putting

$$G = W_G t \text{ and } S = W_S t \qquad (3.34)$$

where W_G = average photon counting rate from glint and W_S = average photon counting rate from speckle and where G refers to the glint and S to the speckle,

$$p(n,t) = \frac{S^n}{(1+S)^{n+1}} L_n\left[-\frac{G}{S(1+S)}\right] \exp\left[-\frac{G}{1+S}\right] \qquad (3.35)$$

where L_n are the Laguerre polynomials.

Tables 3.4 and 3.5 illustrate the effect that different laser radar models have on the probability of detection for an SNR of 20 dB with a probability of false alarm of 10^{-7}, for a coherent system, and a false alarm of 10^{-6} for an incoherent system. For the coherent case, the probability of detection can degrade from 0.9999^+ to 0.85, and for the incoherent case from $0.99.99^+$ to 0.50. These wide variations require the system designer to assess the detection model carefully rather than assume that 20 dB is an acceptable value. Figure 3.24 illustrates that for a Gaussian noise-limited receiver and a diffuse target model, 40 dB of SNR is required for the incoherent detection system, as opposed to 20 dB for the coherent system, in order to achieve a probability of detection of 0.9 with a probability of false alarm of 10^{-7}.

Table 3.4
Laser Radar Models (Coherent Detection) SNR = 20 dB $\cdot P_{FA} = 10^{-7}$

Target Models/Receiver Noise	P_D
• Gaussian Noise	
— Swerling II (Diffuse)	
— No Turbulence	0.85
— Turbulence $\sigma_x^2 = 0.05$	0.85
— Swerling V (Specular)	
— No Turbulence	0.9999+
— Turbulence $\sigma_x^2 = 0.05$	0.95
— Rician	
— Specular + Diffuse (Specular 4× Diffuse)	0.85

Table 3.5
Laser Radar Models (Incoherent Detection) SNR = 20 dB $\cdot P_{FA} = 10^{-6}$

Target Models/Receiver Noise	P_D
• Gaussian Noise	
— Diffuse Target	0.50
— Glint Target	0.9999+
• Photon Noise (Poisson Noise—1 Noise Photon, 10 Signal Photons)	
— Glint Target	0.77
— Diffuse Target	
— Number of Speckles	
1	0.50
4	0.60
Infinite	0.75

Chapter 4
Lasers

4.1 INTRODUCTION

Laser action has been observed at a wide number of wavelengths, covering the optical spectrum from the UV through the far IR; there are even lasers that operate in the microwave region. Neon lasers have operating wavelengths in the UV, argon lasers have wavelengths in the blue or green region of light, and helium neon lasers operate in the red region of light; such a laser can be purchased for approximately $100 per milliwatt of power. Additionally, the ruby (chromium aluminum oxide) laser operates in the red region of the visible spectrum. Semiconductor lasers such as gallium arsenide and gallium arsenide phosphide emit radiation in the near-IR and visible spectra, respectively. Neodymium-type lasers operate at 1.06 μm; such units have been used for accurate range measurements and as designators for precision-guided ordnance. Progressing to the mid- and far-IR portions of the electromagnetic spectrum, the lasers are chemical, operating with hydrogen fluorine (2.8 μm), deuterium fluorine (3.8 to 4.3 μm), carbon monoxide (5 μm), or carbon dioxide (10.6 μm). At the 10.6-μm wavelength, almost any aspect of microwave system operation may be attained because of the high power and good coherence capabilities of these transmitters.

The laser shown in Figure 4.1 has a cavity which looks similar to a fluorescent bulb. On one end of the cavity is a fully reflecting mirror; on the other is a partially reflecting mirror. Coupled between these two mirrors is an active medium—a ruby rod, as in a ruby laser; a gas, such as CO_2; or a semiconductive material, such as gallium arsenide. Basically, this material is one that absorbs energy. Energy is coupled into this active medium by means of a pumping medium. The energy could be coupled into the active medium absorption band via electrical, RF, light, or chemical energy. Assume, for example, an active medium that absorbs 0.53 μm of energy (green light). Because green light is at a higher frequency than red light, the green region of the optical spectrum has more energy. The pumping energy applied to the active medium results in excitation of the molecules; if enough energy is coupled into this rod from the light source, lasing action occurs—all photons within the active laser medium move at the same rate, forming a narrow coherent beam of energy.

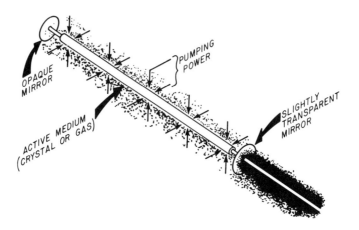

Figure 4.1 Basic laser elements, ©IEEE. (Courtesy of IEEE.)

In the case of the ruby chromium aluminum laser, the chromium absorbs the green. That green light may be obtained from an electrically excited flash lamp which elevates the energy of the atoms in the ground state of the rod, making them "climb up an atom hill." When the atoms get to the top of this "atom hill" (inverted population), some drop back to the "ground." Returning to the ground indicates a change of states—the molecules have changed energy states from one of high energy to a region of low. As a result, an energy difference exists. If the top of the atom hill is thought of as the region of green light and the ground state is conceptualized as the level to which the atoms would like to drop, the difference between the height of the atom hill and the ground represents an energy increment. As these atoms drop to the ground they emit radiation proportional to the difference between the two energy states. For the case of the chromium aluminum laser, the radiation emitted is a 0.6943-μm red light. Descriptions of this process are shown in Figures 4.2 and 4.3. The excited energy causes the atoms to climb the atom hill to position (3).

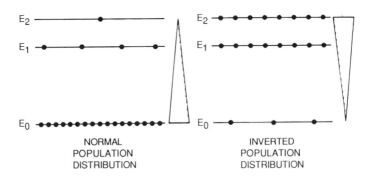

Figure 4.2 Illustration of population inversion.

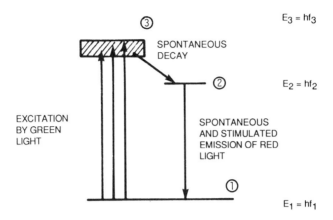

Figure 4.3 Simplified energy-level diagram of chromium atoms in ruby.

Figure 4.4 Neodymium YAG laser. (Courtesy of Raytheon Laser Advanced Development Center.)

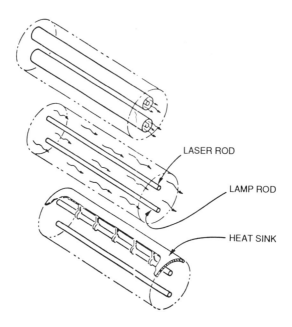

Figure 4.5 Rod cooling techniques.

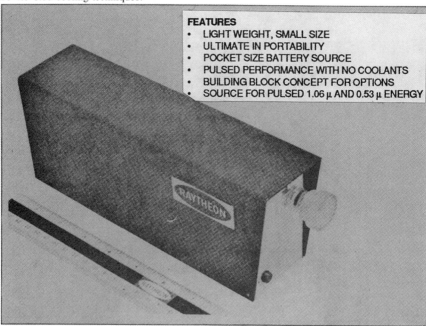

Figure 4.6 Raytheon Model SS-219 miniaturized pulsed YAG laser transmitter.

Some of the energy drops down to energy state (2); it subsequently falls to the ground state (1), resulting in the emission of red light. Basically, what has happened is that the normal population distribution of the material has been inverted so that there are more atoms in an elevated state than there are in a ground state.

Figure 4.4 illustrates a neodymium YAG laser. It is approximately 7 in long, 3 in wide, and 3 in high. It has a heat exchanger which uses a fan to blow air through radiating fins, thereby cooling the liquid surrounding the flash lamp and laser rod (see Fig. 4.5). This type of laser and power supply is typically used in rangefinder applications. Typical characteristics for such a device are shown in Table 4.1.

Recently, improvements in machinery, process, and control technology in the semiconductor industry required for production of VHSIC (very high-speed integrated circuits) and MMIC (monolithic microwave integrated circuits) have been applied to the development of gallium arsenide lasers, resulting in increased efficiency of gallium arsenide-based sources and a substantial increase in output power per unit area. Figure 4.7 illustrates one such Lincoln Laboratory device, where multiple diodes are arrayed on a silicon wafer and the diode array edge surfaces are

Table 4.1
Model SS-219 Miniaturized Pulsed YAG Transmitter

Mechanical Specifications	
Total Transmitter (less accessories)	
Size	$2'' \times 3.25'' \times 7$
Weight (24 Vdc input)	3 lb
(115 Vac 50/60 Hz input)	5 lb
Enclosure	Dustproof
Laser Head (less accessories)	
Size	$1.125'' \times 1.25'' \times 6.75''$
Weight	10 oz
Performance Specifications	
Laser Output (Normal Model)	Threshold to 100 mJ (adjustable)
Pulse Duration	at 1.06 μm = 100 μs
Laser Output (Q-Switched)	50 mJ (nominal) at 1.06 μm
Q-Switched Pulse Duration	10–20 ns
Pulse Repetition Rate	
(Q-Switched and Normal Mode)	1 pps (max.) continuous
Trigger	Manual auto rep. (1 pps) external (+15V, 10 μs)
Laser Crystal	Nd^{-3} YAG 4×50 mm
Polarizer	Calcite glan prism
E/O Q-Switch	$LiNbO_3$ pockels cell
E.M.I.	Unit shielded to minimize electromagnetic interference

Table 4.1 (cont'd)
Model SS-219 Miniaturized Pulsed YAG Transmitter

Electrical Input
DC Option:
  Voltage   24 Vdc (nominal) 18–32V
  Current   2.7A (nominal) 2.2A–3.3A
  Charge Time   0.7 sec (nominal) 0.35–0.9 sec
Battery Pak   250 laser shots between recharges
AC Option:
  Voltage   115 Vac 50/60 Hz
  Power   25W (nominal)

Available Accessories
115-Vac Adapter for 24-Vdc Unit
Pockels Cell Q-Switch
Rechargeable Battery Pak
Frequency Doubler
Boresighting Telescope
Beam Shaping Telescope

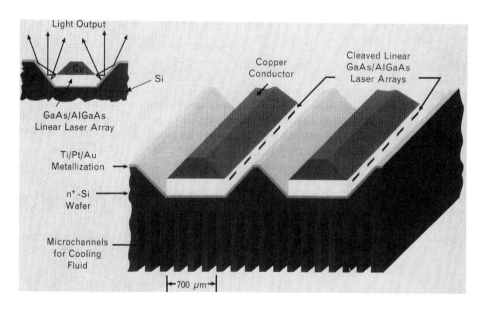

Figure 4.7 Hybrid 2-D surface-emitting array of GaAs/AlGaAs diode lasers with integral Si heat sink. (Reprinted, by permission, from The Lincoln Laboratory Journal [75].)

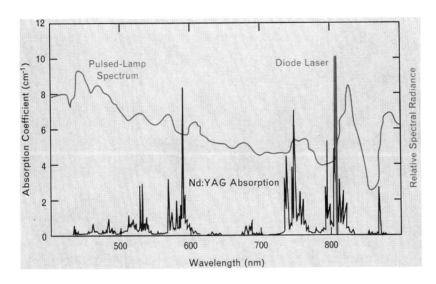

Figure 4.8 Absorption spectrum of Nd:YAG emission spectra of a diode laser and a pulsed flashlamp. The absorption spectrum is 1%-doped Nd: YAG. (Reprinted, by permission, from The Lincoln Laboratory Journal [75].)

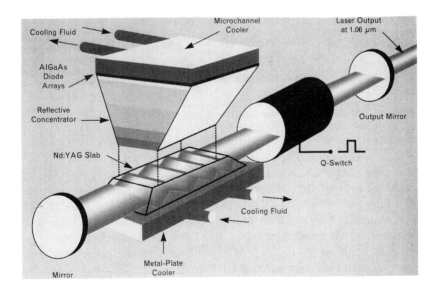

Figure 4.9 Diode-laser-pumped zig-zag slab laser concept. (Reprinted, by permission, from The Lincoln Laboratory Journal [75].)

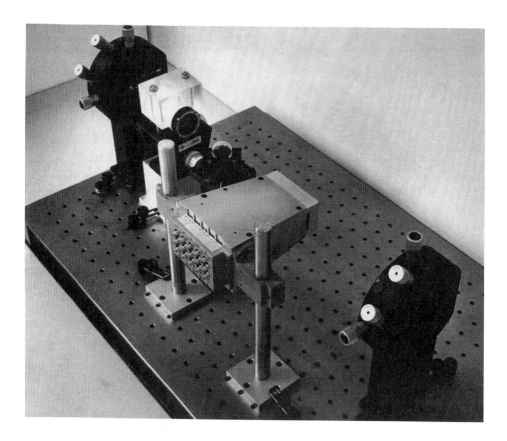

Figure 4.10 Ten-watt diode-laser-pumped zig-zag slab laser. (Reprinted, by permission, from The Lincoln Laboratory Journal [75].)

configured into a reflecting surface to allow the output power of each diode source to be extracted over an area composed of many sources. The resulting high power density in turn can be used as a laser source or as a flashlamp substitute to pump other laser materials. Figure 4.8 illustrates the pulsed flashlamp and diode laser emission spectrum and the neodymium (Nd):YAG absorption spectrum.

Efficient pumping of the Nd:YAG can occur with the diode laser source because most its radiation may be coupled into an absorption band of the Nd:YAG laser, thereby generating less heat in the rod and resulting in higher output power, repetition rate, and coherence. Figure 4.9 illustrates that increasing the surface area of the Nd:YAG material via a zig-zag pattern can allow more diode array pumping surface to be available, thereby increasing the pumping power. Figures 4.10 and 4.11 illustrate the zig-zag slab laser and the output power obtainable from such a

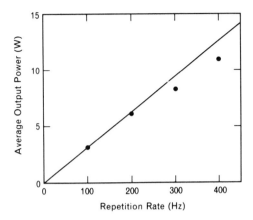

Figure 4.11 Average power output from the zig-zag slab laser as a function of the pulse repetition rate. The straight line is a fit to the low repetition-rate data points. The data points at a high repetition rate fall below this line because of power supply limitations. (Reprinted, by permission, from The Lincoln Laboratory Journal [75].)

Table 4.2
Compiled from the 1991 Laser Focus World Buyer's Guide

	Wavelength (μm)	Laser	Power/CW (Watts)	Pulsed Energy (Joules)	PRF (Hertz)	Pulse-width (microseconds)
1	0.157	Fluoride		30	25	10
2	0.19–4.5	Dye laser		0.1	2–40	0.004
3	0.193	Argon fluoride		0.2	50	0.17
4	0.22	Krypton chloride		100	50	20
5	0.248	Krypton fluoride		500	200	20
6	0.263	Nd: glass		0.8	0.03	0.01
7	0.266	Neodymium (Nd): YAG		0.1	20	0.01
8	0.308	Xenon chloride		0.5	300	0.03
9	0.337	Nitrogen		0.001	100	0.0003
10	0.35	Nd: glass		2.2	0.03	0.01
11	0.351	Xenon fluoride		300	500	30

Table 4.2 (cont'd)
Compiled from the 1991 Laser Focus World Buyer's Guide

12	0.45–0.51	Argon	20			
13	0.51	Copper		8K	0.02	19
14	0.527	Nd: glass		22	0.008	0.02
15	0.53	Nd: YAG	10–20	0.75	10	0.04
16	0.63	Semiconductor	0.003			
17	0.67	Ti: sapphire	0.7			
18	0.67	Semiconductor	5			
19	0.68–1.07	Ti: sapphire	0.5	0.3	1–20	0.01
20	0.7	Alexandrite		2.5	5	200K
21	0.72–0.78	Alexandrite		1	1–20	0.05
22	0.75	Semiconductor	0.003			
23	0.77–0.84	Semiconductor	0.1	12K	100	20
24	0.78	Semiconductor	1.0			
26	0.79	Semiconductor	0.003	10		
27	.808	Semiconductor	10	200K	100	200
28	0.88	Semiconductor	10		1000	0.1
29	1.04–1.2	Alexandrite		0.1	10–30	0.05
30	1.054	Nd: glass	1–1000	100	0.008	0.02
31	1.06	Nd: YAG		1000	1500	500
32	1.15	Helium neon	0.001			
33	1.27–1.33	Semiconductor	0.002			
34	1.315	Atomic iodine		3	10	5
35	1.318	Nd: YAG				
36	1.52–1.58	Semiconductor	1			
37	1.72–2.5	CO: MGF$_2$		.020	20	80
38	2.1	Holmium: YAG	0.8	2	1–20	200
39	2.6–3	Hydrogen fluoride	1000			
40	2.940	Erbium: YAG		1	50	200
41	2.940	Erbium		1	10	250
42	3.3–27	Semiconductor	0.001			
43	3.3–4.2	Semiconductor	0.0003			
44	3.6–4	Deuterium fluoride	100	0.03	25	0.15
45	4.2–6	Semiconductor	0.001			
46	5–6.5	Carbon monoxide	10			
47	6–8.5	Semiconductor	0.001			
48	8.5–10	Semiconductor	0.0001			
49	9–11	Carbon dioxide	1000	5000	50K	0.01

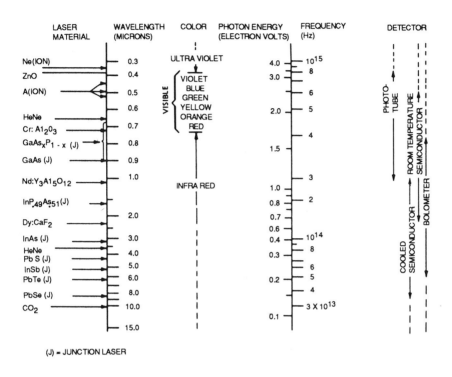

Figure 4.12 User's guide to lasers. (*RCA Electro-Optics Handbook* [4].)

transmitter capability. Because the Nd:YAG absorption band is narrow at the gallium arsenide pump band, the diode laser must be thermally controlled to prevent the diode laser from shifting its frequency outside the peak absorption band and degrading the overall efficiency.

Table 4.2 illustrates some of the more popular laser sources, and Figure 4.12 presents a user's guide to lasers.

4.2 DETECTORS

When a detector operates in a square law region, the power incident upon the detector results in a current at the output directly related to the incident power; this current may be expressed as

$$i = \rho_i E^2 \tag{4.1}$$

where

ρ_i = detector responsivity (amps/watt).

The electric field, E, incident upon the detector may be expressed as

$$E = A_l \cos\omega_s t \tag{4.2}$$

The detector output current, i, can be shown to be

$$i = \rho_i A_l^2 (1/2 + 1/2 \cos 2\omega_{st}) \tag{4.3}$$

For an optical detector, ρ_i relates to the conversion of optical energy to electrical energy and is termed the detector responsivity.

4.2.1 Detector Responsivity

For ideal optical detection, one photon of light releases one electron from the detector. The number of photons, M_p, required to emit these electrons for the nonideal detector then becomes a measure of how efficiently the detector operates and is typically termed the device quantum efficiency, η.

$$M_p = \frac{1}{\eta} \tag{4.4}$$

The current produced by M_p photons per second may be expressed as

$$i = \eta M_p q \tag{4.5}$$

where q = electron charge.

As indicated before, the energy level associated with each photon is expressed as the product of Planck's constant, h, and transmitting frequency, f.

Correspondingly, the number of photons per second may be expressed as

$$M_p = \frac{P_{SIG}}{hf} \tag{4.6}$$

Therefore, the detector current is

$$i = \frac{\eta_D q P_{SIG}}{hf} \tag{4.7}$$

Thus ρ_i, the detector current responsivity, is equal to $\eta_D q/hf$. The detector current responsivity, ρ_i, physically represents the ratio of the RMS current out of the detector to the RMS power incident onto the detector; it has units of amperes/watt. Correspondingly, the voltage responsivity, ρ_v, has units of volts/watt.

4.2.2 Noise Equivalent Power

The noise equivalent power (NEP) may be expressed as

$$\text{NEP} = \frac{\text{Detector Noise Current}}{\text{Responsivity}} \tag{4.8}$$

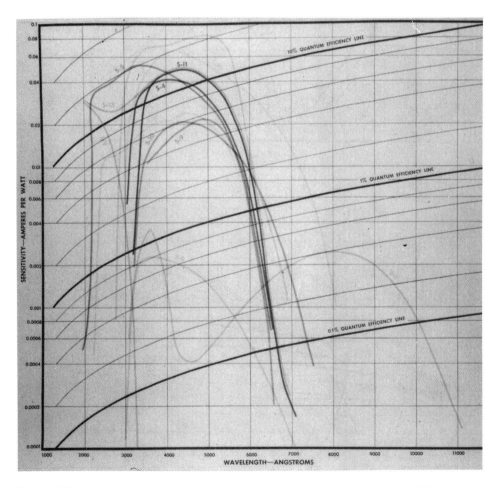

Figure 4.13 Typical absolute spectral response characteristics of photoemissive devices [76].

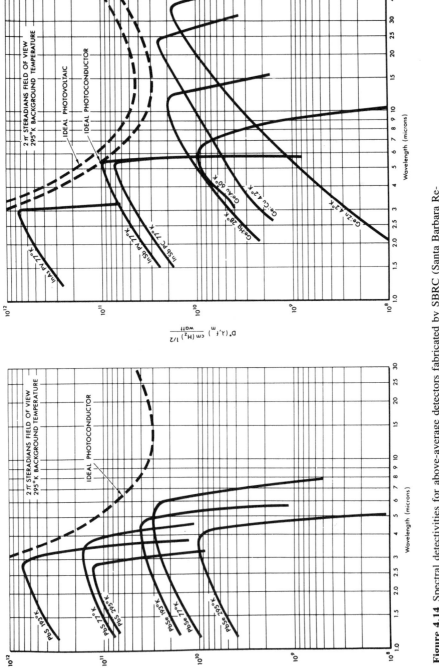

Figure 4.14 Spectral detectivities for above-average detectors fabricated by SBRC (Santa Barbara Research Center of Hughes Aircraft Company). A reduction in background photon flux produces higher detectivities [43].

and represents the amount of RMS modulated power applied to the detector (an amount equal to the RMS detector noise voltage), usually specified as

NEP (500K, 1000, 1)
Radiation Measurement Detected
Source Frequency (Hz) Bandwidth (Hz)
Temperature

Because NEP is dependent on the area of the detector, the term $D*$ was developed to standardize detector comparisons; this term references detector measurements to a 1-cm² area, in a bandwidth of 1 Hz.

$$D* = \frac{(A\,\Delta f)^{1/2}}{\text{NEP}} \frac{\text{cm} - (\text{Hz})^{1/2}}{\text{W}} \quad (4.9)$$

Often, $D*$ is specified as $D*(\lambda, f_m)$ where f_m is the modulating frequency in hertz referred to a 1-Hz receiver bandwidth.

Note that $D*$ is also dependent on the signal wavelength. Figures 4.13 [76] and 4.14 [43] and Table 4.3 indicate typical detector sensitivities and characteristics throughout the visible, near-, mid-, and far-infrared wavebands [77].

Table 4.3
Detector Characteristics

Photoemissive Detectors				
Spectral Range (μm)	0.145–0.87	0.145–1.1	0.4–1.2	
Responsivity (μA/μW @ λ)	128 @ 0.4	18 @ 1.06	0.0019 @ 0.8	
Dark Current (A)	5×10^{-10}	3×10^{-10}	3×10^{-7}	
Quantum Efficiency (%)	15	2	0.2 @ 0.9	
Rise and Fall Times (ns)	0.32	0.12	1.5	
Photocathode Type	S-20	InGaAs P	AgOC$_s$	
Pyroelectric Detectors				
Spectral Range (μm)	0.001–1000	0.2–1000	2.5–30	0.2–500
Peak Detectivity (cm-Hz$^{1/2}$/W)	2×10^6	10^7	10^8	10^8
Responsivity (μV/μW @ λ)	200 @ 10	0.3×10^{-6} (μA/μW)	10^{-6} A/W	10^{-5}–100 V/W
Rise Times (sec)	10^{-7}	10^{-8}	6×10^{-8}	5×10^{-4}–5×10^3
Type	LiTaO$_6$	Ferroelectric	PVF$_2$	SBN
Dimensions	1–5 mm	0.5–4 mm	10 mm	3–20 mm

Table 4.3 (cont'd)
Detector Characteristics

Semiconductor Detectors (Visible–Near IR)

Spectral Range (μm)	0.3–1.8	0.4–1.1	0.35–1.13	0.4–1.1
NEP (W) or Detectivity (cm-Hz$^{1/2}$/W)	10^{-16}	4.5×10^{-13}	6×10^{11}	5×10^{-15}
Responsivity μA or μV μW @ λ	0.7 μA/μW	0.6 @ 0.9	0.5 @ 0.9	110 μA/μW
Peak Wavelength (μm)	1.5	0.9	0.95	0.9
Rise Time (μs)	10^{-2}	0.12	0.01	0.02
Type	Ge	Silicon	Silicon P1N	Silicon avalanche
Dimensions	0.8 mm^2	11-mm diameter	11-mm diameter	3.2-mm diameter

Semiconductor Detectors (Mid IR)

Detector Type	Indium antimonide (PV)	Lead sulfide (PC)	Lead selenide (PC)	Lead selenide (PC)
Temperature (K)	77 or 243	77	195	300
Spectral Response	0.5–6 μm	0.5–5 μm	0.5–6 μm	0.5–4.7 μm
D* (cm-Hz$^{1/2}$/W)	$1–3 \times 10^{11}$	$1.5–2.5 \times 10^{11}$	2×10^{10}	1.5×10^9
Peak Wavelength	5 μm	2.8 μm	4.8 μm	3.8 μm
Responsivity	1.5–2 A/W	0.1–1 A/W	0.01–0.05 A/W	$4–30 \times 10^{-3}$ A/W
Rise Time (μs)	0.02–0.2	$2–5 \times 10^3$	10–40	1–3
Active Area (mm)^2	0.008–7	0.01–100	0.01–100	0.01–100

Semiconductor Detectors (Far IR)

Detector Type	Lead tin telluride (PV)	Mercury cadmium telluride (PC)	Mercury cadmium telluride (PV)
Temperature (K)	77	77	77 or 243
Spectral Response	0.5–14	0.5–13	2–12
D* (cm-Hz$^{1/2}$/W)	2×10^{10}	$1–4 \times 10^{10}$	$2–5 \times 10^9$
Peak Wavelength	10	10.6	10.6
Responsivity	3 A/W	3–30 A/W	1.3–4.3 A/W
Rise Time (μs)	1–2	0.2–0.8	$1.6–16 \times 10^{-4}$
Active Area (mm)^2	0.008–3	$2.5 \times 10^{-3}–9$	0.008–0.8

Chapter 5
Incoherent Receiver Detection Systems and Techniques

Figure 5.1 illustrates a functional block diagram of a laser rangefinder. A source of electrical energy is used to charge up an energy storage network which, upon application of a triggering signal, transfers this energy to a laser. The energy coupling effectively into the laser results in a pulse of light energy being emitted from the laser and propagated through an optical system that provides the desired transmitter beam divergence to the target.

Mounted within the optical cavity is a high-bandwidth optical diode that detects the output laser energy and provides a reference timing signal to the receiver electronics and elapsed time counter. Energy backscattered from the target then passes through suitable collection optics to a detector. Standard threshold detection circuitry, range gating, and elapsed time counting are used to provide accurate ranging and range rate information. Because of the short pulsewidth and narrow beamwidth associated with this type of optical system, precision target range information can be obtained.

Figure 5.2 illustrates an artist's concept of the laser altimeter system used on Apollos 15, 16, and 17 to provide accurate ranging information for the camera system flown on these flights. As a result of having accurate ranging information to a known point within the camera field, accurate scale factoring and significant improvement in map accuracy were obtained. Equipment specifications for this system are noted in Table 5.1. Figure 5.3 shows the hardware, built by RCA in Burlington, Massachusetts.

Figure 5.4 illustrates an AN/GVS-5 hand-held laser rangefinder [78] developed by RCA for the U.S. Army and having the specifications shown.

Similar transmitters operating at higher energy and repetition rate can be used to illuminate and spot-designate targets. Figure 5.5 illustrates the AN/TVQ-2 U.S. Army ground laser locator designator (GLLD), developed by the Hughes Aircraft Company. This man-portable unit precisely locates and designates targets for attack

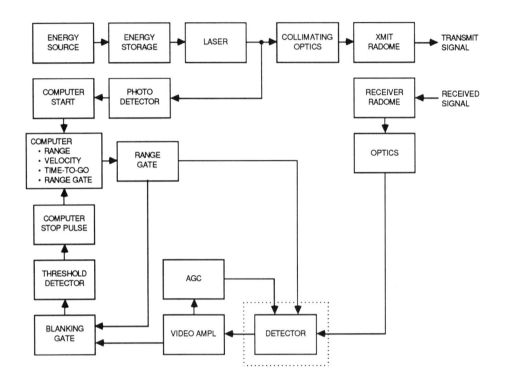

Figure 5.1 Functional block diagram of laser rangefinder.

by armed helicopter, laser homing weapons, or conventional artillery fire. Figure 5.6 illustrates a Copperhead cannon-launched guided projectile (CLGP), developed for the U.S. Army by the Martin Marietta Aerospace/Orlando Division. Upon firing, movable fins are deployed and the projectile is roll-rate stabilized. After a preset time the projectile changes from a ballistic trajectory to a guided one when the guidance sensor, located in the tip of the projectile, acquires the laser-reflected energy from the target being designated by the GLLD. Guidance signals are derived from this reflected energy to allow the movable fins to guide the GLLD to the target, which may be located from 3 to 16 km away. Figure 5.6 illustrates a CLGP approaching and impacting a test target.

Correspondingly, target-reflected signals from a laser target designator may be collected by the USAF Pave Penny pod mounted on an A-10 aircraft boresighted pylon (Fig. 5.7). Figure 5.8 shows the laser-reflected energy from the designated target on a head-up display. The laser-reflected energy is denoted by the diamond symbol.

Figure 5.2 Apollo laser altimeter.

Table 5.1
Laser Altimeter Equipment Specifications (Courtesy of RCA.)

Laser
Q-switched ruby—mechanical Q-switch
Energy output >250 mJ/pulse
Pulse repetition frequency (PRF): 1 pulse per 15 sec maximum
"Solo" mode—1 per 20 sec
"Camera" mode—slaved to 3-in metric mapping
  camera: 1 per 15 sec or slower
  typical at 60 nmi altitude
  1 per 25 sec
Beam divergence: 80% of output energy in 0.3 mrad
  (Collimating telescope 16 × 4 inches clear aperture, catadioptric)

Table 5.1 (cont'd)
Laser Altimeter Equipment Specifications (Courtesy of RCA.)

Receiver
Extended red S-20 photomultiplier detector
25Å bandwidth ca 6943Å
Field of view: 0.2 mrad

Altitude Measurement
Minimum altitude: 40 nmi
Maximum altitude: 80 nmi
Resolution: 1m
Accuracy: ±2m

Altitude Output
18-bit binary to telemetry
Difference format for mapping camera—recorded on each film frame
Also analog and digital status outputs to telemetry

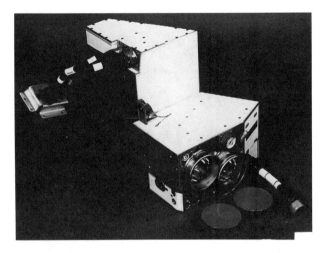

Figure 5.3 Apollo laser altimeter hardware (Courtesy of RCA.)

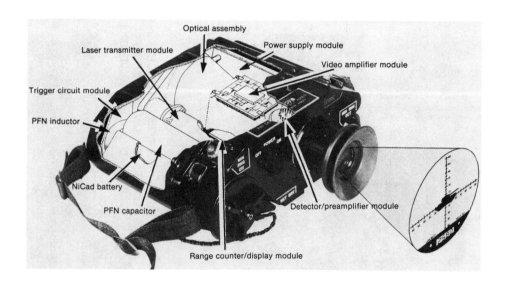

Maximum range	9990 m
Minimum range	≤200 m
Range resolution	±10 m
Range accuracy (R.S.S.)	10 m
Minimum-range gate continuous adjustment range	≤200 m to ≥5000 m
Sight field of view	7°
Resolution	7 arc-sec., on axis
Receiver field of view	1 mil
Receiver bandwidth	20 nm
Receiver sensitivity	2 nW/cm^2
False-alarm rate (internal noise)	<0.01
Transmitter power	2 MW in 1.0-mil beam
Transmitter pulse width	6 ns
Ranging duty cycle	96/hr (continuous)
Recycle time	1 s
Ranges per battery charge	400–700
Weight	≤4.7 lb. w/battery

Figure 5.4 AN/GVS-5 hand-held laser rangefinder. (Courtesy of RCA.)

Figure 5.5 AN/TVQ-2 U.S. Army ground laser locator designator (GLLD). (Courtesy of Hughes Aircraft Company.)

LOAD
CONFIGURATION

FLIGHT
CONFIGURATION

Figure 5.6 Copperhead cannon-launched guided projectile (CLGP). (Courtesy of Martin-Marietta Aerospace.)

Figure 5.7 Pave Penny pod.

Table 5.2 [4] compares the relative performance of direct detection (incoherent) ruby and neodymium YAG lasers for a range of 5 km. Here, the ruby system is evaluated using an S-20 photo cathode surface and a silicon semiconductor detector. Similarly, the neodymium system is evaluated for an S-1 photo surface photomultiplier and a silicon detector. The sun irradiance and the optical filter bandpass is twice as large for the ruby system than it is for the YAG. As a result, the background power, P_{BK}, is four times larger.

Evaluation of the noise terms indicated that the silicon detector systems are receiver thermally noise limited; correspondingly, the photomultiplier systems are signal shot noise limited, with the background noise power approximately an order of magnitude or more below the signal shot noise level. The incoherent detection signal-to-noise ratio (SNR) equation (1.55) indicates where further effort should be concentrated in order to improve the system's performance; Table 5.3 illustrates the

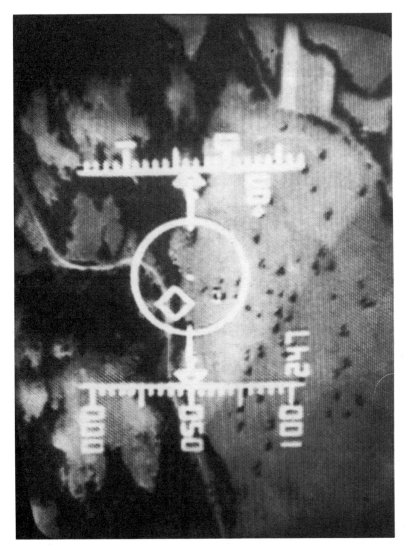

Figure 5.8 Laser-reflected energy from designated target on a head-up display. When laser energy reflected from the designated target is received by Pave Penny, the exact target location is represented on the head-up display by the diamond symbol. (Courtesy of Martin-Marietta Co.)

Table 5.2
Summary of Parameters and Performance Calculations for Four Laser Rangefinders (Courtesy of RCA [4].)

		Ruby		Neodymium	
	Units	S-20	Si	S-1	Si
Range, R	km	5 →			
Wavelength, λ	μm	0.694 →		1.06 →	
Attenuation Coefficient, σ	km^{-1}	0.139 →		0.114 →	
Atmospheric Transmittance, T_a	—	0.449 →		0.565 →	
Sun Spectral Irradiance, $H_{\lambda s}$	W m^{-2} Å^{-1}	0.12 →		0.06 →	
Filter Leakage Transmittance, X	—	0 →			
Backscatter Coefficient, σ_s	km^{-1}	0 →			
Background Power, P_{BK}	W	1.66×10^{-10} →		4.6×10^{11} →	
Pulsewidth, τ	ns	20 →			
Bandwidth, B	MHz	25 →			
Optical Filter Bandpass, B_o	Å	40 →		20 →	
Receiver Lens Diameter, d_r	in	2.8 →			
Geometry Factor, M	—	1 →			
Receiver Beamwidth, α_r	mrad	1 →			
Transmitter Beamwidth, α_t	mrad	1 →			
Target Reflectance, ρ	—	0.1 →			
Receiver Noise Factor, F	—	1.5 →			
Detector Load Resistor, R_L	Ω	1000 →			
Transmittance, receiver optics, T_r	—	0.7 →			
Transmittance, transmitter optics, T_t	—	0.7 →			
Target Incidence Angle, θ	deg	0 →			
Signal-to-Noise Ratio, SNR	—	53 →			
Detector Gain, G	—	5×10^4	1	1.5×10^5	1
Cathode Dark Current, I_D	pA	0.030	10^5	12.9	10^5
Detector Responsivity, β	A W^{-1}	0.028	0.517	3.5×10^{-4}	0.152
Peak Received Signal Power, P_s	W	1.23×10^{-9}	2.5×10^{-7}	1.64×10^{-7}	8.48×10^{-7}
Single Pulse Range Accuracy, δR	m	0.82	0.82	0.82	0.82

Table 5.3
Calculated Peak Laser Power to Range of 5 km in a Clear Standard Atmosphere for Four Laser Rangefinders [4]

Laser Material	Detector	Peak Laser Power Required, P_t (kW)
(1) Ruby	S-20 photomultiplier	2.6
(2) Ruby	Silicon photodiode	405
(3) Neodymium YAG	S-1 photomultiplier	482
(4) Neodymium YAG	Silicon photodiode	1,120

Reprinted, by permission, from *RCA Electro-Optics Handbook*

transmitter power requirements for each system. Tables 5.4 and 5.5 and Figure 5.9 illustrate the configuration and specifications of an incoherent detection YAG laser radar system developed by Sylvania for cooperative target tracking.

Figure 5.10 illustrates the first U.S. 10.6-μm laser rangefinder, which has improved adverse weather and smoke penetration capability over visual and near-IR wavelength systems, and is also eye safe. Figure 5.11 illustrates some performance specifications of this unit, which was built by Raytheon Company, Equipment Division, Wayland, Massachusetts.

Table 5.4
Sylvania 1.06-μm Laser Radar

Transmitter	Q-spoiled Nd:YAG
Peak Power	10^6 W
PRF	100 pps
Beamwidth	10 mr
Pulsewidth	25 ns
Range Measurement Accuracy	±1/2 foot
Detector Quantum Efficiency	>50%
Receiver	Quadrant photodiode
Angle Tracking (Azimuth and Elevation) accuracy	±0.1 mrad
Target	3 in retroreflector
Range	10 nmi
Initial Acquisition	TV

(Courtesy of Sylvania).

Table 5.5
PATS Incoherent Detection YAG Laser Radar, Based on
use of 3-in-Diameter 4-arc-sec Retroreflector

1. Absolute Accuracy*
   Azimuth: ±0.01% of range (0.1 mrad) (for target ranges of 500 to 65,000 ft)
   Elevation: ±0.01% of range (0.1 mrad) (for target ranges of 500 to 65,000 ft)
   Range: ±1 ft for target ranges of 700 to 30,000 ft
   ±2 ft for target ranges of 30,000 to 65,000 ft
2. Maximum Range 100,000 ft
   Accuracy for target at 100,000 ft
   a. Azimuth: ±0.3 mrad   b. Elevation: ±0.3 mrad   c. Range: ±5 ft
3. Data Sample Rate: 100, 50, 20, or 10 sample sets selectable
4. Angular Coverage
   Azimuth: ±170° (multiple turn capability available)
   Elevation: −5° to +85° (dynamic specifications apply for elevation angles between −5° and +45°)
5. Acquisition Dynamics (manual—using joystick and TV monitor)
   Maximum angular rate (azimuth and elevation): 100 mrad/s
   Maximum angular acceleration (azimuth and elevation): 80 mrad/s^2
6. Acquisition Dynamics (automatic-aided with 100 sample/sec coordinate data)
   Maximum angular rate (azimuth and elevation): 500 mrad/s
   Maximum angular acceleration (azimuth and elevation): 80 mrad/s^2
7. Autotrack Dynamics
   Maximum angular rate (azimuth and elevation): 500 mrad/s
   Maximum angular acceleration (azimuth and elevation): 80 mrad/s^2
8. Operator Displays
   Range: Digital in 1-ft increments Azimuth: Digital in 1° increments
   Elevation: Digital in 1° increments
9. Acquisition Field of View: 5° to 20° with zoom control
10. Viewfinder Field of View: 3 mrad
11. Power Requirements: 208 V, 3-phase, per electrical performance specifications of MIL-STD-633
12. Environmental Conditions (Operating):
    Ambient temperature: −20° to 120°F   Wind: 0 to 50 knots
13. Setup Time: <1 hour

*After computer smoothing, a weighted averaging technique is used with a sliding window of 0.1 sec and a sample rate of 100/s.
(Courtesy of Sylvania.)

Figure 5.9 Sylvania incoherent detection YAG (1.06 μm) laser radar used for cooperative target tracking.

150

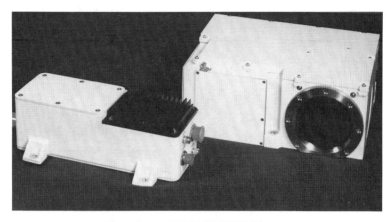

10.6 MICRON LASER RANGEFINDER

RANGE	5 KM (FLIR COMPATIBLE)
RANGE ACCURACY	± 10 METERS
PRF	5 PPS
MULTIPLE TARGET	FIRST PULSE/LAST PULSE
SAFETY	EYE SAFE

Figure 5.10 First U.S. CO_2 laser rangefinder by Raytheon Company, Equipment Division, Sudbury, Massachusetts.

RANGE (CLEAR)	5 KM
BEAM DIVERGENCE	0.5 MR
PULSE WIDTH	40 NS
RANGE ACCURACY	$<\pm$ 10 METERS
PRF	1 TO 5 PPS
MULTIPLE TARGET	FIRST PULSE/LAST PULSE
DATA OUTPUT	BCD SERIAL
BUILT IN TEST EQUIPMENT	DETECTOR READY T_{ZERO} READY RECEIVER READY EXTERNAL VIDEO
COMMAND AND CONTROL	PRF 1 PPS/5 PPS FIRST PULSE LAST PULSE FIRE/NO FIRE STATUS FLIR SYNC
SAFETY	EYE SAFE
SIZE	13.2 × 10 × 5.75 12 × 6 × 4
WEIGHT	31.75 POUNDS; 8.75 POUNDS

Figure 5.11 10.6-μm laser rangefinder parameters.

Chapter 6
Coherent Laser Systems and Techniques

6.1 INTRODUCTION

Laser radar systems employing coherent detection receivers have been used in field and flight test experimentation since 1968. A generic laser radar block diagram is shown in Figure 6.1.

In this configuration the laser is modulated to provide information to the transmitted signal, which is coupled through the interferometer, optics, and scanner to illuminate the scan field of interest. The received signal is then coupled, via reciprocity through the interferometer, to the receiver detector, where it is mixed with a sample of the laser signal in the form of a local oscillator.

The receiver output is processed by the signal processor to extract target information and then processed by the data processor where all information is compiled to provide target position, range, velocity, and an image. Figure 6.2 illustrates a 1968 ground signal returned from a Raytheon 5W coherent 10.6-μm airborne system. Figures 6.2(A) and 6.2(B) illustrate this Doppler-shifted signal along with an aerial camera photograph with the laser boresighted to the center of the frame. Figure 6.2(C) illustrates the Doppler-shifted signal of a pedestrian walking a crosswalk, as observed in the aerial camera photograph to the left. In the next sequence, the laser footprint passes over the pedestrian and the aircraft ground speed is observed in pictures 6.2(D) and 6.2(E). Early success of these experimental systems encouraged further technology development. However, laser devices were too large, weighing as much as 450 lb. Subsequently, Raytheon developed an air-cooled 5W laser weighing 7.5 lb, including power supplies and cooling. The coherent 5W laser was configured into the scanning laser radar shown in Figure 6.3. This system was designed, fabricated and flight tested under USAF and DARPA sponsorship in 1975. A conceptual diagram of the system is shown in Figure 6.3. Here the 0.5-mrad beam was propagated to the terrain via a Palmer scan. The beam advances in a contiguous manner across the terrain (via aircraft flight velocity) to yield a coherent Doppler spectrum in the receiver.

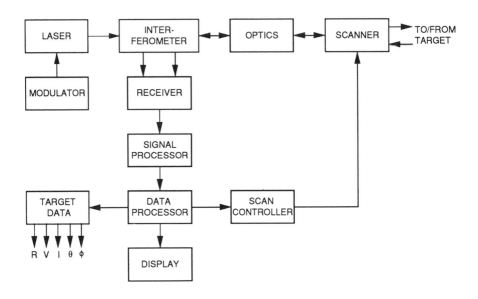

Figure 6.1 Typical laser radar block diagram.

The backscattered Doppler-shifted target signal was then processed in a surface acoustic wave signal processor, recorded on tape, and subsequently played back on the ground through a CRT display. The system was configured for a remotely piloted vehicle (RPV) application with the tape recorder simulating the data link. Thresholding of the Doppler-processed signal intensity resulted in gray scale rendition on a photographed CRT. Figure 6.4 illustrates a strip made in such a manner. The CO_2 laser radar map, made at a 50° depression angle, is shown as the central image in Figure 6.4. Surrounding this picture are 70-mm aerial camera photographs from a vertically oriented camera. It can be readily observed that optical-quality photographic images may be made with laser radars, and also that speckle effects may be reduced by suitable processing.

Having illustrated optical resolution capability, let us now evaluate the MTI radar aspects of this system. Figure 6.5 illustrates a 70-mm aerial camera scene of a forested area at Fort Devens, Massachusetts. There is a tank suitably noted in Figure 6.5(A) traveling on a road. A return signal from the forest or road is Doppler shifted relative to the aircraft velocity and appears at a frequency $f_o + f_D$, while a return signal from the moving tank appears at a frequency $f_o + f_D + f$ target and, as such, appears in a different Doppler filter. The surface acoustic wave delay line processor is programmed to give an MTI cue, in real time, when a Doppler return

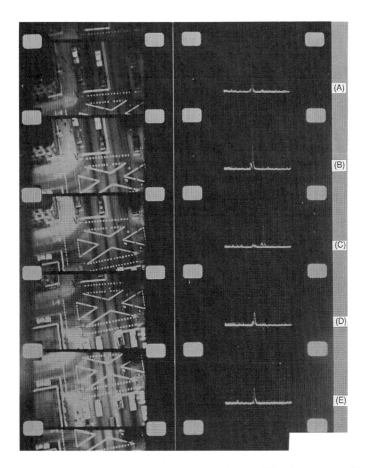

Figure 6.2 CO_2 laser Doppler airborne moving target indication (AMTI) of a person walking across an intersection (McManus, Chabot, Goldstein, Delgrego).

greater than 5 knots is indicated in adjacent pixels. Figure 6.5(B) illustrates this moving target flag, which occurs at a full intensity gray shade on the display. Additionally, it may be noted that several target clutter intensity spots cause false alarm indications which are distributed through the scene. Altering the target detection algorithms to require an "N detections out of M trials" detection approach results in Figure 6.5(C) and a significant false alarm reduction. Also note what appears to be a second target in this scene, which was not observed in the photograph. A clutter patch of the duration and continuity of motion, required to pass the MTI target detection requirements, tends to rule out a false target detection.

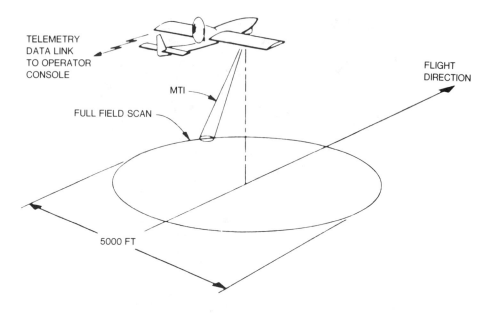

Figure 6.3 RPV Palmer scan geometry (Keene, Chabot, McManus, Delgrego).

Forward-looking ground-based systems [79] have been configured to demonstrate target acquisition and imaging capabilities. Figure 6.6 illustrates MTI detection and imaging over a 6° × 4° raster scan field of a moving tank, obtained by a Triservice Raytheon system, and shows a photograph of an M-48 tank and CO_2 laser radar image of the same vehicle. Tank targets tended to have a significant diffuse component, as shown by the number of consecutive returns. This may be contrasted to that of a helicopter shown in Figure 6.7, where specularities contributed to the image. Helicopter blade rotation caused the Doppler return to be outside the bandwidth of the imaging filter so that the blades were not imaged.

Figure 6.8 illustrates an amplitude-modulated (AM) CW laser received signal from a stationary target which has been processed to illustrate range measurement characteristics [79]. Typically, color is used to display the range images. In this image, the range ambiguity of the 8-MHz AM tone is seen as horizontal bands over the display. Normally, the color scale bar under the picture is utilized to indicate range increments over the ambiguity intervals (i.e., returns at longer ranges are one color while shorter ranges are a different color) and can be used to highlight the tank turret and gun barrel. Figure 6.9 illustrates how edge extraction techniques implemented in the computer can be used to highlight internal features of another vehicle. Here, the hood area, cab, and rear body are highlighted.

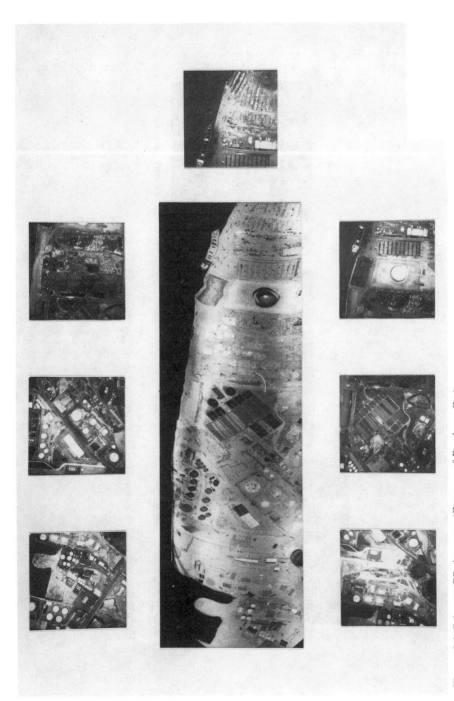

Figure 6.4 Coherent CO_2 imagery. (Courtesy of Raytheon Co.)

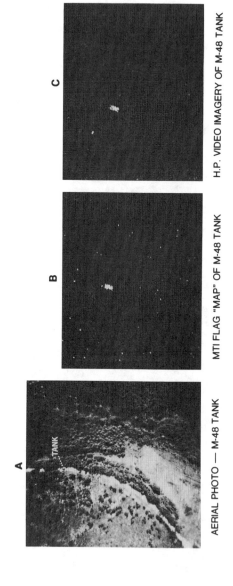

Figure 6.5 AMTI detection of M-48 tank at altitude of 1000 ft. (Courtesy of Raytheon Co.)

RANGE	250 M
VELOCITY	10 MPH
S/N	~25 dB
BW	100 kHz
VIDEO PROC	LOGARITHMIC

Figure 6.6 M-48 tank approaching. (Courtesy of Raytheon Co.)

Simultaneous measurement programs with passive IR systems have also been conducted. One of these results indicates the diurnal washout of a passive 8–12 μm thermal image of a tanker truck located at a range of 4 km, along with that of a CO_2 laser range image of the same vehicle. Because the range image is not dependent on time of day and thermal contrast changes in backgrounds and targets, reliable signatures are obtained from laser radar systems. Figure 6.10 also illustrates the ability of the active system to measure wires.

In 1984, Raytheon proposed the use of these laser radar principles to the USAF Armament Division, Eglin AFB, Florida, for the purpose of autonomous guidance for tactical stand-off weapons against—

- Mobile SAM sites
- Bridges
- Buildings
- Runways
- Buried targets

System requirements necessitated simultaneous active and passive operations at 10.6 and 8 to 12 μm, respectively. This hardware was designed and supplied to the USAF for a captive flight demonstration in 1987. The purpose of the captive flight test was to demonstrate—

- Autonomous, real time, target detection, and classification
- Autonomous aimpoint selection

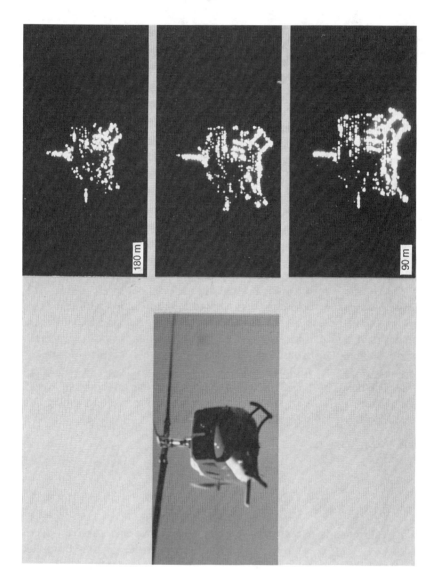

Figure 6.7 UH-1 helicopter. (Courtesy of Raytheon Co.)

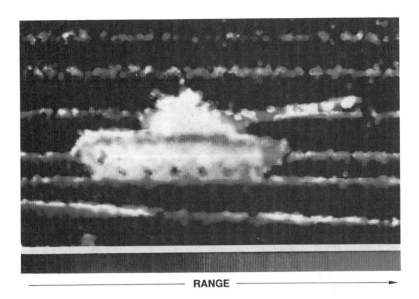

Figure 6.8 Laser radar range image. (Courtesy of Raytheon Co.)

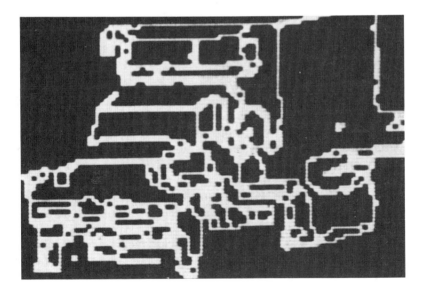

STATIONARY TARGET
INDICATION (STI)

Figure 6.9 Range image edge extinction. (Courtesy of Raytheon Co.)

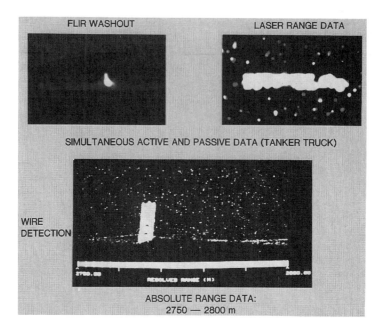

Figure 6.10 CO_2 laser radar capabilities. (Courtesy of Raytheon Co.)

Figure 6.11 illustrates the seeker hardware mounted in the nose section of a C-45 aircraft, along with photographs of the bridges and tactical target scenes.

Figure 6.12 illustrates simultaneous active and passive airborne bridge and ship imagery, while Figure 6.13 illustrates the AMTI target detection of cars and trucks on a highway, along with passive thermal imaging of the same scene showing the targets and the roadway. As the system was not roll-stabilized, the highway does not appear as a straight line. In Figure 6.14, active and passive imagery of an airport scene may be observed. Here, aircraft parked in front of the hangar area have different characteristics in two scenes. The range image illustrates two parked aircraft, while the passive scene suggests three. This is due to a thermal shadow caused by a recently departed aircraft.

A video TV camera was boresighted to the sensor and a computer and graphics generator was used to indicate target detection via a box generated on the detected target. Real-time classification of the target resulted in the suitable detection box overplayed on the target location. Figure 6.15 illustrates the bridge detection box located over several pier structures. Subsequent real-time computer processing resulted in the computer selection of a bridge pier aimpoint. Similar detection, classification, and aimpoint selection occurred for mobile targets.

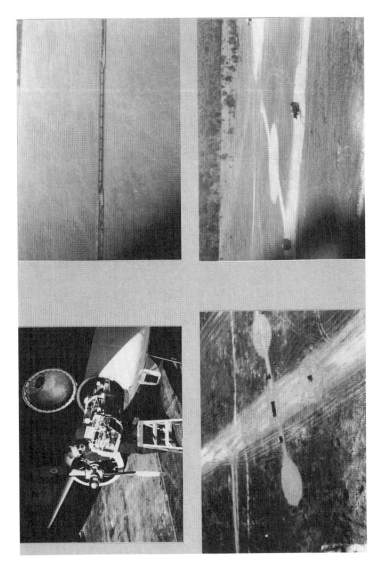

Figure 6.11 Flight tests. (Courtesy of Raytheon Co.)

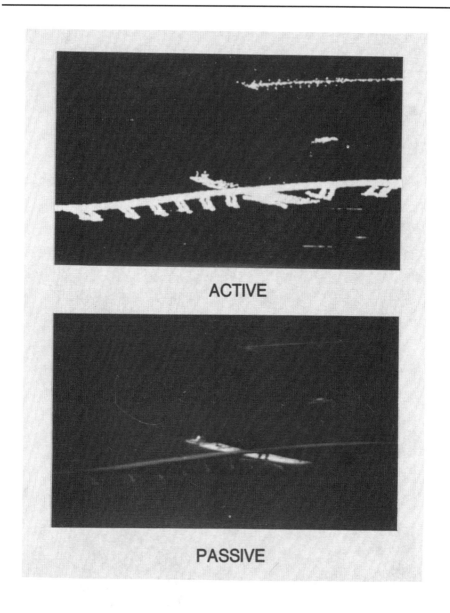

Figure 6.12 Active and passive bridge images. (Courtesy of Raytheon Co.)

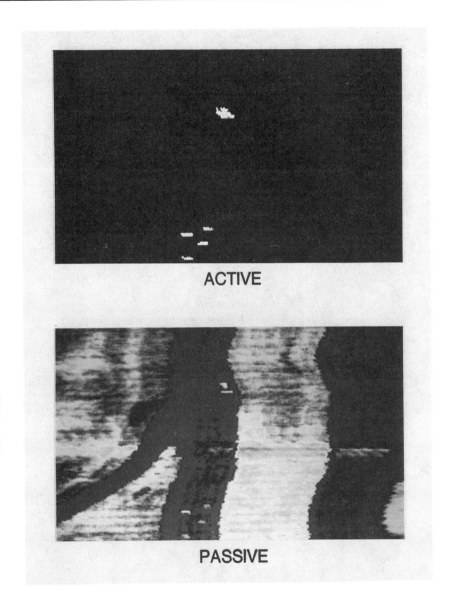

Figure 6.13 Active and passive images—MTI detection. (Courtesy of Raytheon Co.)

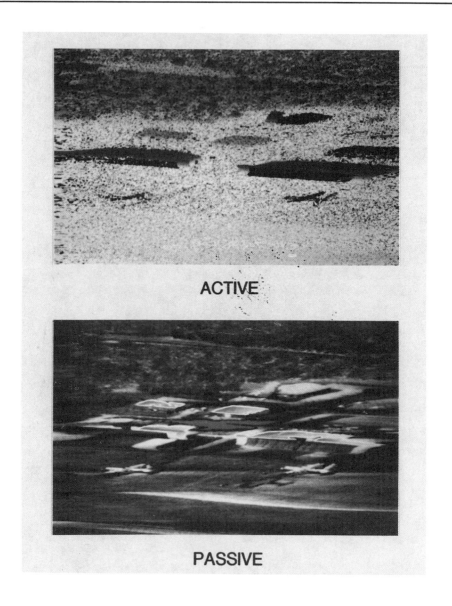

Figure 6.14 Active and passive images—hangar complex. (Courtesy of Raytheon Co.)

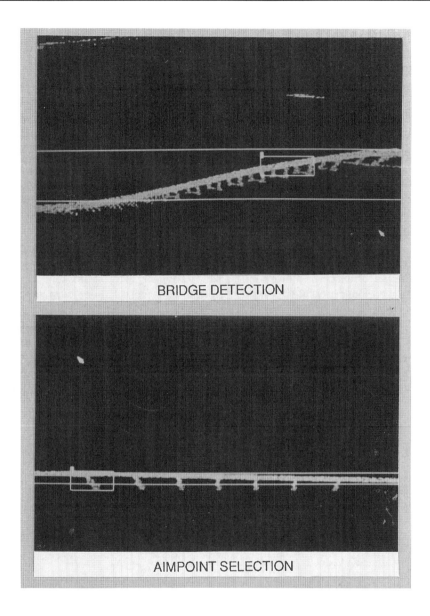

Figure 6.15 Bridge detection and aimpoint selection (active). (Courtesy of Raytheon Co.)

6.2 FIRE CONTROL LASER RADAR ADJUNCTS

Laser radar systems can operate as adjuncts to microwave radar or passive IR systems. In this role, limited data rate laser radars can be used to scan the region containing the potential target. Here, the target angular position, range, and velocity may be determined more accurately to potentially improve the false target reporting of the primary sensor. The following evaluates a 10.6-μm and a 1.06-μm laser radar configuration for this role.

In order to demonstrate the capability of coherent detection systems such as these for fire control applications, calculations were performed using the parameters described in Table 6.1. Range performance as a function of atmospheric conditions is portrayed in Figure 6.16 for a target with a cross section of 1m^2. Even in hazy conditions the system has greater than a 10-km capability, while in 4 mm/hr of rainfall the performance calculation indicates a greater than 5-km capability. An equivalent comparison of a coherent 40 millijoule 1.06-μm system is also illustrated; here, the performance in atmospheric haze and fog is seen to substantially degrade system performance, while the rain performance is very similar between the two wavelengths.

A similar set of calculations were conducted [80] for a 100-mJ, 1.06-μm laser having an incoherent receiver, with the performance parameters shown in Figure 6.17. Ranges in excess of 10 km may be realized from this system in clear weather. This may be observed in Figure 6.18, which shows the probability of detection *versus* SNR as a function of range with a neodymium laser. Under the conditions of

Table 6.1
CO_2 Laser System Parameters

Transmitter Peak Power	10 kW
Optics Diameter	10 cm
Transmitter/Receiver Beamwidth	0.1 mrad
Pulse Width	1.0 to 10 μs
	Nominal 4 μs
Detector Quantum Efficiency	0.5
Detector Temperature	77K
System Losses	20 dB
SNR	20 dB
Target Cross Section	1.0m^2
Detection Technique	Coherent
Operating Wavelength	10.6 μm
Repetition Rate	Handover volume dependent

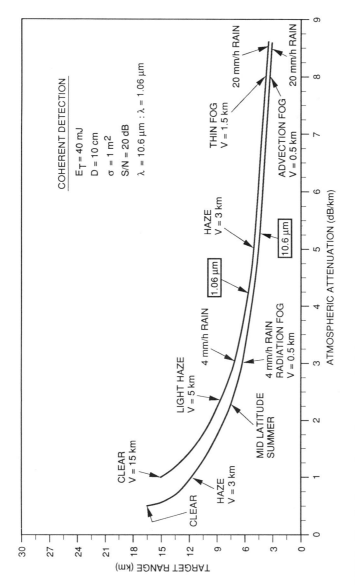

Figure 6.16 Pulse Doppler laser radar performance calculations.

YAG[A] LASER SYSTEM PARAMETERS

LASER:

WAVELENGTH	1.06 μm
PULSE ENERGY	100 mJ
PULSE LENGTH	0.02 μs
BEAM DIVERGENCE	0.4 mrad
CROSS SECTION	1 m^2
S/N	22 dB
RANGE[B], CLEAR (VISIBILITY ~15 nmi)	5.7 nmi (10.4 km)
RANGE[B], HAZE (VISIBILITY ~2 nmi)	2.5 nmi (4.6 km)

NOTES: (A) WITH QUADRANT SILICON AVALANCHE PHOTODIODE DETECTOR

(B) AGAINST A 1 m^2 (σ) TARGET FOR INITIAL DETECTION, DETECTION PROBABILITY = 95% PER SCAN

Figure 6.17 YAG[A] laser system parameters.

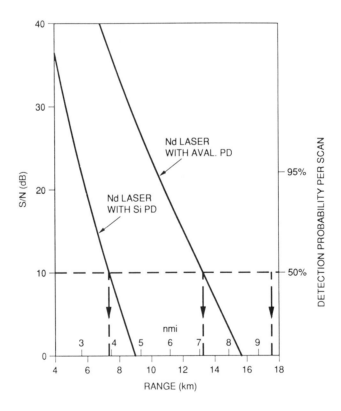

Figure 6.18 Sensitivity of Nd:YAG laser system in clear weather (visibility 30 km).

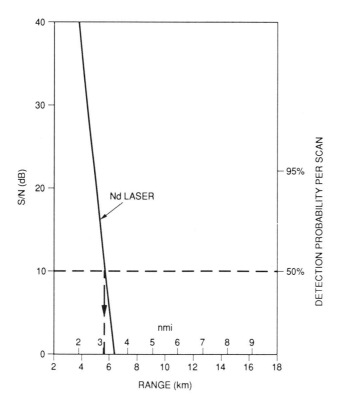

Figure 6.19 Sensitivity of Nd:YAG laser system visibility—4 km (haze).

clear visibility, a 50% probability of detection may be observed in excess of 7 nmi for the neodymium laser with silicon avalanche detector, and approximately 4 nmi for a silicon detector. In the event of haze, the incoherent YAG system range performance (Figure 6.19) is reduced to less than 3 km.

6.2.1 Long-Range Laser Radars

Figures 6.20 and 6.21 and Table 6.2 [81,82] illustrate the ground-based Lincoln Laboratory Firepond 10.6-μm laser Doppler radar facility and its block diagram and specifications for the system as it existed around 1975.

This equipment was used at a reduced power level to track the GEOS-III satellite at an approximate range of 1100 km, with the coherent optical monopulse receiver illustrated in Figure 6.22. Initial target acquisition was accomplished with the Millstone Hill radar with the aid of a low light level TV tracker. Elevation angle

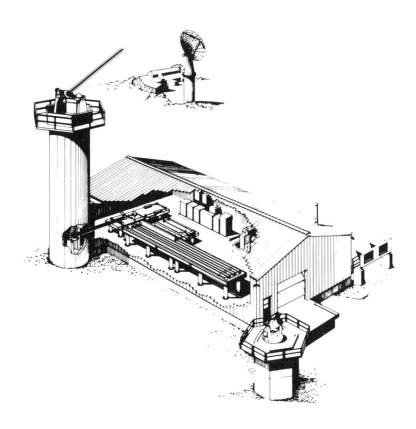

Figure 6.20 Lincoln Laboratory Firepond 10.6-μm Doppler radar facility. (Courtesy of MIT Lincoln Laboratory.)

Table 6.2
Lincoln Laboratory Firepond 10.6-μm Laser Doppler Radar
(Courtesy of R. Kingston and T. Gilmartin; MIT Lincoln Laboratory.)

Wavelength	10.6 μm
Configuration	MOPA
Transmitter Power	15 kW peak, 1.4 kW average
Pulse Repetition Frequency (PRF)	10 kHz
Laser Frequency	3×10^{13} Hz
Laser Stability	20 Hz over 50 msec
Telescope	48 cm; f/7 Cassegrain
Offset Laser Frequency	5 MHz
Beat Frequency Difference Between Two Lasers	<1 kHz over seconds
Visual Track	Wide-angle TV
IR Angle Track	Conical scan amplitude—monopulse
Detector	Cu:Ge
Detector Bandwidth	1.2 GHz (15,000 mph)

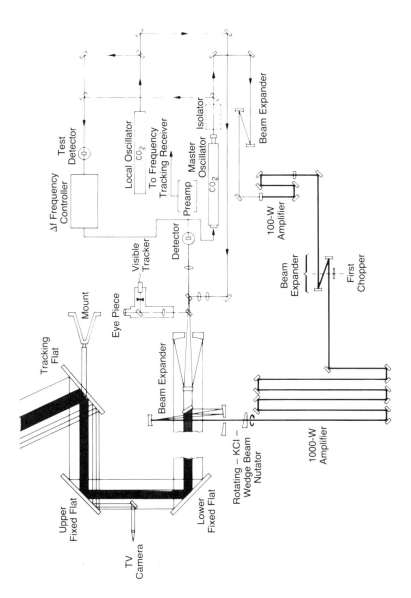

Figure 6.21 Lincoln Laboratory Firepond CO$_2$ laser radar schematic. (Courtesy of MIT Lincoln Laboratory.)

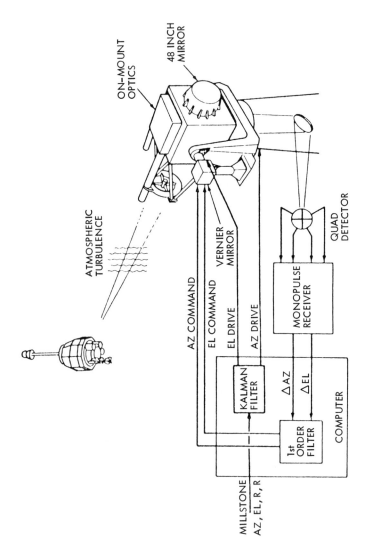

Figure 6.22 Satellite tracking mode (long-range mode) of Lincoln Laboratory Firepond 10.6-μm laser radar.

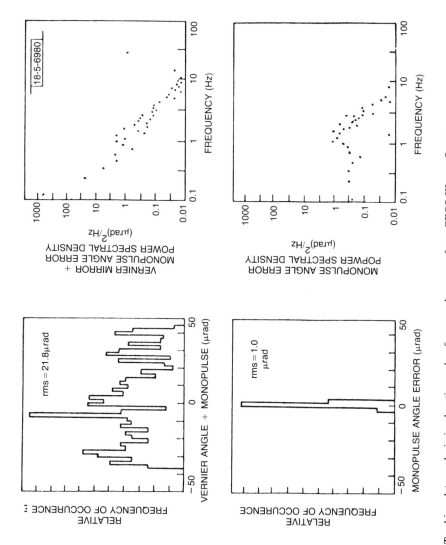

Figure 6.23 Tracking data analysis in elevation only of monopulse errors from GEOS-III retroreflector at range of about 1000 km.

Object #	Name	Type	Size (meters)	Track ID	Minimum Range (km)	Comment Spectrum	Comment Monopulse Track
5679	OPS 7898 RB	Agena RB	6.6 x 1.5	80029.0016	1100	Slightly broadened	Yes
4954	OPS 5268 RB	Burner II	1.7 x 1.6	80051.2322	925	Narrow	Yes
2825	OPS 5712 RB	Agena D	6.1 x 1.5	80052.0103	1160	Slightly broadened	Yes
8820	LAGEOS	Sphere, Retros	0.5	80052.0234	5900	Narrow but cyclic frequency excursions	Yes
2825	OPS 5712 RB	Agena D	6.1 x 1.5	80056.2355	1100	Slightly broadened	Yes
5398	LCS IV	Al. Sphere	1	80057.0035	1160	Narrow	Yes
4954	OPS 5268 RB	Burner II	1.7 x 1.6	80057.2355	840	Slightly broadened	Yes
5398	LCS IV	Al. Sphere	1	80058.0021	1050	Narrow	Yes
8820	LAGEOS	Sphere, Retros	0.5	80058.0127	6020	Narrow but cyclic frequency excursions	Yes
5398	LCS IV	Al. Sphere	1	80059.2353	970	Narrow	No
8820	LAGEOS	Sphere. Retros	0.5	80060.0214	5900	Narrow but cyclic frequency excursions	Yes
3892	NIMBUS 3 RB	Agena D	6 x 1.5	80060.0845	1200	Broad & complex	Yes
4332	OPS 0054 RB	Burner II	1.7 x 1.6	80060.0929	880	Slightly broadened	Yes
4954	OPS 5268 RB	Burner II	1.7 x 1.6	80060.1020	920	Slightly broadened	Yes
5556	OPS 4316 RB	Burner II	1.6 x 1.6	80060.1028	920	Extremely broad	No
3892	NIMBUS 3 RB	Agena D	6 x 1.5	80064.0928	1300	Broad & complex	Yes
4332	OPS 0054 RB	Burner II	1.7 x 1.6	80064.0941	860	Slightly broadened	Yes
4954	OPS 5268 RB	Burner II	1.7 x 1.6	80064.1009	1040	Slightly broadened	Yes

Figure 6.24 1980 Firepond tracking results.

tracking statistics are shown in Figure 6.23. Vernier mirror tracking data portraying the difference between the tracking algorithm (Kalman filter) pointing data and the monopulse line of sight indicated that the coarse pointing errors of approximately 100 μrad were tracked out of an rms value of 1 μrad, with a tracker bandwidth of about 1 Hz [83]. Figures 6.24 and 6.25 illustrate some of the Firepond tracking and Doppler spectra results from space objects at distances of approximately 1000 km.

During the 1980s, the Firepond facility was improved through the addition of a wide-bandwidth FM capability having the ability to perform range-Doppler imaging, as well as the use of argon, ruby, and Nd:YAG solid-state lasers, as shown in Figure 6.26. The facility is shown in the foreground of Figure 6.27. In 1990, the system was used to track a rocket launched from Wallops Island, Virginia, in combination with microwave radar systems at Wallops Island and Millstone Hill, Mas-

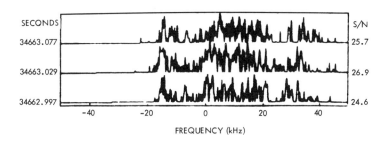

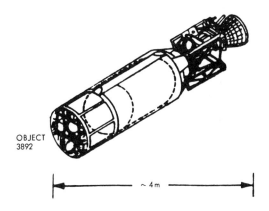

Figure 6.25 Doppler spectra of three IR radar returns.

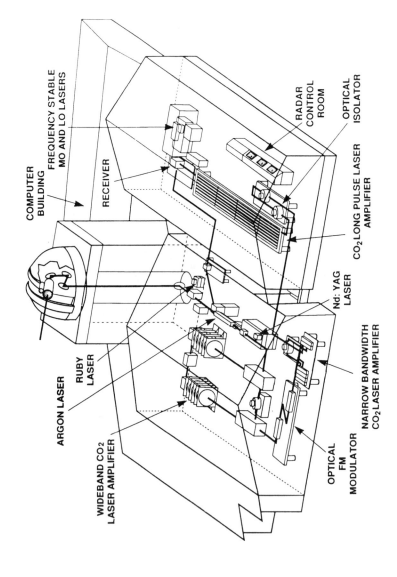

Figure 6.26 Firepond laser radar. (Reprinted by permission of W. Keicher, MIT Lincoln Laboratory.)

Figure 6.27 Millstone Hill site. (Reprinted by permission of W. Keicher, MIT Lincoln Laboratory.)

sachusetts [84,85]. The Firefly II laser experiment is shown in Figure 6.28. In order for the laser radar to acquire, track, and image the target, a complex set of tracking events were necessary.

The NASA Wallops Island C-band radar acquired and tracked the rocket shortly after launch and passed the track data across telephone lines [84] to MIT's Lincoln Laboratory Millstone Hill Field site in Westford, Massachusetts. The Millstone Hill deep space surveillance L-band radar and Haystack X-band imaging radar accepted the data and tracked the rocket. A 24-in telescope at the Firepond laser radar site then acquired and tracked the rocket. This coordination was a significant real-time demonstration of sensor data fusion and cooperation. As the rocket passed above a 20° elevation angle, the Firepond site began illumination of the rocket with an argon ion laser and the high-power isotopic carbon dioxide laser radar. The argon ion laser provided a signal for an angle tracker while the high-power carbon dioxide laser radar performed range-Doppler imaging (also known as inverse synthetic aperture imaging) of the target. The carbon dioxide laser radar uses a telescope with a 48-in aperture. The laser radar imaging of the unenhanced target took place at ranges of 600 to 750 km.

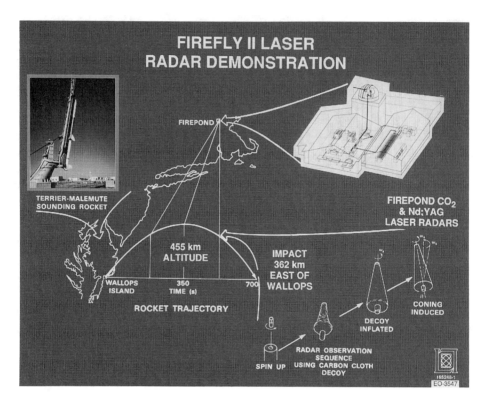

Figure 6.28 Firefly II laser radar demonstration. (Reprinted by permission of W. Keicher, MIT Lincoln Laboratory.)

Chapter 7
Atmospheric Laser Radar Systems

7.1 INTRODUCTION

Laser radar systems have been used to directly measure properties of the atmosphere. These systems use the backscattered energy from illuminated molecules or aerosols normally suspended in air to remotely measure particle concentration, mixing levels, pollution concentration, trace constituents, wind velocity, turbulence, and energy levels of the atmosphere. System operation is similar to those discussed earlier for "hard targets"; however, the "effective target section" (σ) described is replaced with the volumetric cross section associated with the atmosphere. The volumetric effective cross section (σ_V) contains molecules and aerosols having scattering properties determined by Rayleigh and Mie scattering theories. The backscatter properties can be written as a product of the number density (N_z) and cross section of the aerosols and/or molecules. Lawrence, McCormick et al. [86,87] in "Optical Studies of the Atmosphere" showed that the molecular and aerosol scattering coefficients B_M and B_A could be expressed as

The scattering coefficients for the molecular components is $B_M(z)$:

$$B_M(z) = \int_0^{2\pi} \int_0^{\pi} 1/2 \, (1 + \cos^2\theta_{SC}) \, \sigma_{CM}(z) \, \sin\theta_{SC} \, d\theta_{SC} \, d\phi \qquad (7.1)$$

where

$$\sigma_{CM} = \left(\frac{2\pi}{\lambda}\right)^4 \overline{\alpha^2} \, N(z) f_A \qquad (7.2)$$

and

$\bar{\alpha}$ = polarizability
$N(z)$ = the number density at an altitude z
$f_A \sim 1$

The scattering coefficient for the aerosol component, which obeys a Junge distribution, is $B_A(z)$:

179

$$B_A(z) = S' r_1^{S'} N_A(z) \pi \left(\frac{(2\pi)}{\lambda}\right)^{S'-z} K(\alpha, \eta, S') \tag{7.3}$$

where

$$K(\alpha, \eta, S_{AT}) = \int_{\alpha_1}^{\alpha_2} \frac{Q(\alpha, \eta)}{\alpha^{S_{AT}-1}} d\alpha \tag{7.4}$$

$$Q(\alpha, \eta) = \frac{2}{\alpha^2} \sum_{m=1}^{\infty} (2M + 1)[|a_m|^2 + |b_m|^2] \tag{7.5}$$

and

a_m, b_m	=	Mie coefficients
S'	=	size distribution parameter
$N_A(z1)$	=	total aerosol number density
$\alpha = \dfrac{2\pi r}{\lambda}$	=	particle size parameter
r	=	radius of scatterer
$dn(r_1 z)$	=	number density of particles with radius between r and $r + dr$ at altitude z
θ_{SC}	=	scattering angle between the direction of incident and scattered radiation

This model assumes a Junge size distribution law where $r_1 \ll r_2$

It is readily observed from the above that calculations of combined molecular and aerosol backscattering coefficients is a complex task requiring computer simulations and assumptions of an atmospheric model. This model must consider the distribution, composition, size, and shape of the atmospheric scatterers, among other parameters. As a result, various atmospheric models have been used to represent the atmosphere. Two of the earlier popular models were those of Junge, which used size parameters representative of water aerosols and continental haze, and Deirmendjian, which considered size distributions representative of fogs, clouds, and maritime haze.

Figures 7.1 and 7.2 are examples of McCormick's predictions of backscattering coefficient for molecular and aerosol scattering as a function of altitude (z) for the wavelength associated with the ruby laser 0.6943 μm and its frequency-doubled component 0.3472 μm [88]. Figure 7.3 illustrates the backscatter coefficient for the Deirmendjian model as modified by Wright et al. [88].

In order to modify the range equation for use with atmospheric systems operating in the far field, the effective target cross section term σ must be replaced with (σ_v) where

$$\sigma_v = \frac{4\pi}{1} (B(\pi)) \frac{\pi}{4} R^2 \theta_T^2 \Delta L \tag{7.6}$$

where $B(\pi)$ is the atmospheric backscattering function in units of M^{-1} sr^{-1}.

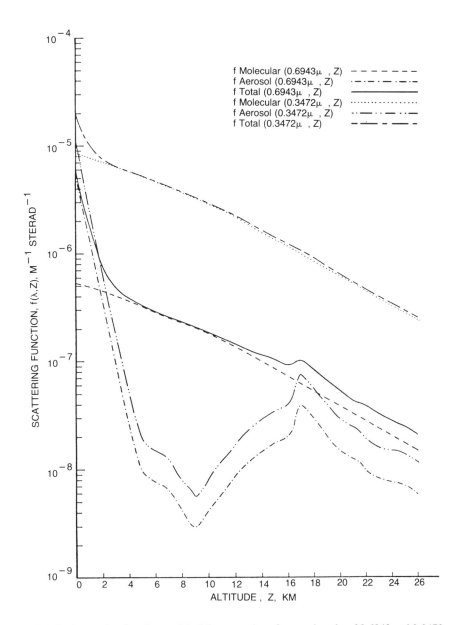

Figure 7.1 Backscattering function model of the atmosphere for wavelengths of 0.6943 and 0.3472 μm. The molecular contribution, aerosol contribution, and their sum are shown.

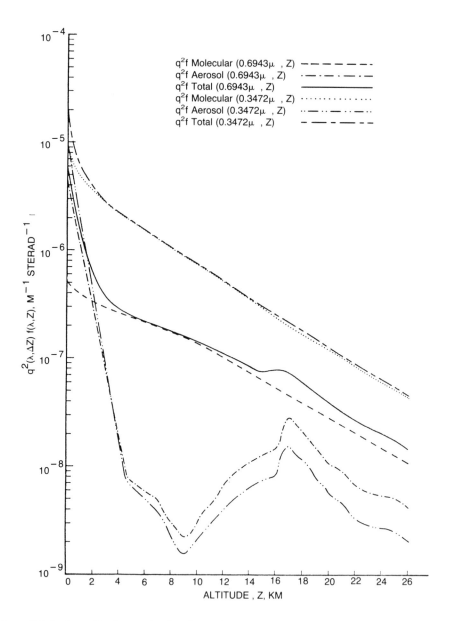

Figure 7.2 Product of backscattering function and two-way transmissivity as a function of altitude for wavelengths of 0.69435 and 0.3472 μm.

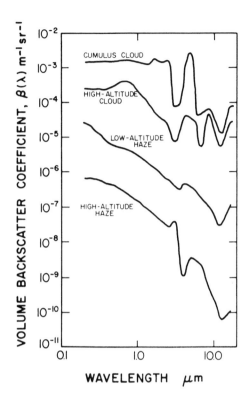

Figure 7.3 Aerosol volume backscattering coefficient as a function of wavelength [88].

For a pulse system, the length of the atmospheric volume sampled is $c\tau/2$, where c is the speed of light in meters per second. The effective volumetric scattering function to be substituted in the range equation becomes

$$\sigma_v = \frac{\pi^2}{2} B(\pi) R^2 \theta_T^2 c\tau \tag{7.7}$$

7.2 REMOTE ATMOSPHERIC MEASUREMENT TECHNIQUES

Both incoherent and coherent detection techniques can be used to provide remote measurements of the atmosphere.

7.2.1 Incoherent Systems Approaches

- *Direct Detection*
  Single-frequency direct detection uses the backscattered signal strength from molecules and aerosols to provide a measure of particle concentration.
- *Absorption*
  The laser radiation is selected to match an absorption band of a given molecule, and the attenuation of the radiation is measured.
- *Fluorescence*
  Laser frequency is chosen such that the atoms or molecules absorb the incident radiation, resulting in a fluorescent emission that may be detected by the receiver.
- *Differential Absorption*
  In this technique, molecular absorption of gases H_2O, O_2, O_3, CO_2, *et cetera*, in the atmosphere is utilized by tuning a laser to a trace gas absorption line and measuring the magnitude of the backscattered energy as well as that of a second frequency in an adjacent window, which is not absorbed by the trace gas, and is used as a reference comparison signal. The difference in the absorption between the absorbed and nonabsorbed beams can then be used to measure the distribution of absorption gases.
- *Raman Scattering*
  In general, the radiation that is backscattered by molecules in the atmosphere is returned at the same wavelength (excluding Doppler motion) as that with which it was illuminated. Raman scattering occurs when the incident radiation transfers energy to the molecule, thereby exciting vibrational states and resulting in the scattering of energy from the molecule at a wavelength different from the original radiation. This wavelength emission from the molecules can be used to determine the atmospheric constituency. Pulse energy propagated into the atmosphere can be range resolved and the wavelengths analyzed to determine remotely chemical composition of the atmosphere.

This subject is covered comprehensively in *Laser Remote Sensing* by Raymond Measures [89], and so will not be covered in great detail here. Rather, an indication of the types of systems and measurements representative of the technology will be discussed.

7.2.2 Incoherent Systems

In a series of papers [86,87,90], the NASA Langley team of McCormick, Lawrence, and Melfi illustrated the ability of incoherent systems to perform most of the above measurements by using an airborne ruby laser radar system and a ground-based multiwavelength (0.34, 0.503, 0.69, and 1.06 μm) coupled through a 48-in telescope (Figure 7.4).

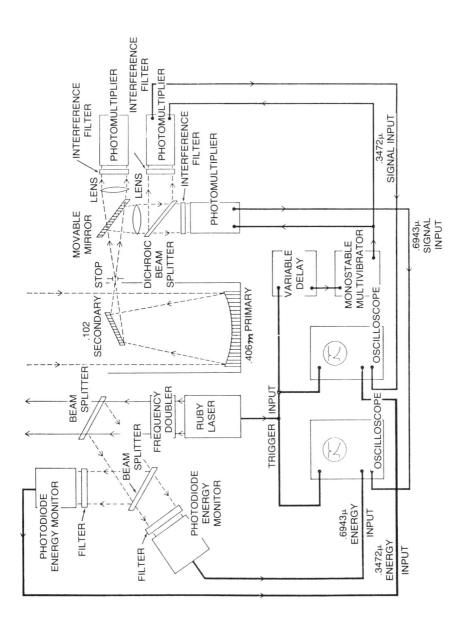

Figure 7.4 Block diagram of the experimental arrangement.

In these systems, the laser energy is propagated into the atmosphere which, because of its transmission and volumetric backscatter coefficient, supplies a continuous return signal to the coaxial receiver as the pulse propagates through the atmosphere. Initially the receiver is gated off to prevent receiver saturation. Upon application of a receiver gate turning the detector/electronics on, the atmospheric signal may be observed at some time t_1 at a range R_1. The energy from the range-resolved window ($C\tau/2$), consistent with a matched filter receiver, is then observed in the time domain to the range and sensitivity limits of the receiver. When viewed on an intensity-time display, the signal has large returns at short range, which decays with time depending on the range dependency, atmospheric transmission, and other factors.

Figure 7.5 [86] illustrates such a display. Calibration of the system sensitivity with time allows the measurement of the backscattering coefficient through knowledge of the system parameters and "ground truth" information. Detailed understanding of the system losses are required in order to achieve a good absolute measure. Figure 7.6 [86] illustrates the relative volume backscatter cross section for aerosols

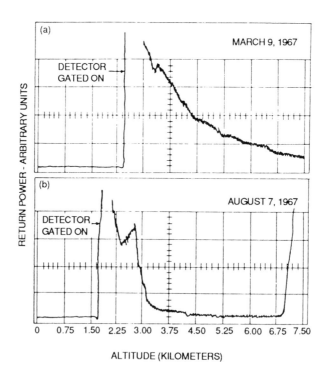

Figure 7.5 Atmospheric intensity/time delay.

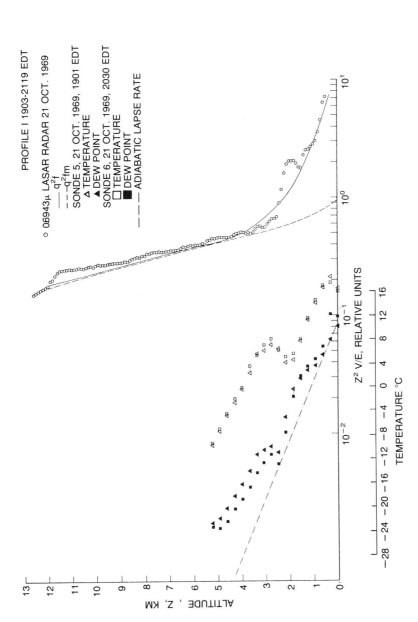

Figure 7.6 Z^2V/E *versus* Z laser radar data taken from 1903 to 2119 EDT on October 21, 1969, for $\lambda = 0.6934$ μm. The data are normalized to the predicted profiles of $q_f^2 f$ in Figure 7.2. Also shown are the temperature profiles from radiosondes launched at 1901 and 2030 EDT on October 21, 1969, and an adiabatic lapse rate curve.

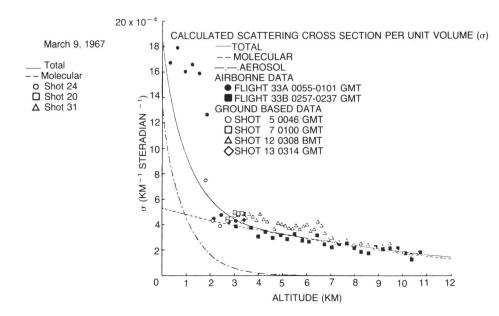

Figure 7.7 Relative volume backscatter cross section for aerosols and molecules as a function of altitude.

and molecules as a function of altitude, while Figure 7.7 shows the measured and theoretical values in units of inverse kilometer-steradians (km^{-1} sr^{-1}) for a pulsed ruby laser system, while measurements made at a wavelength of 0.3472 μm, obtained by harmonic generation techniques, are shown in Figure 7.8.

Figure 7.9 compares the aerosol mixing ratio (W_A) for the ruby laser channel against that indicated by rawinsonde measurements.

The two-wavelength data were used to show that the atmosphere behaved as a molecular scatterer above the mixing layer to 12 km, the aerosol data correlated very well with the water vapor mixing process and that dynamic atmospheric processes could be remotely monitored.

Remote monitoring of the atmosphere with Raman scattering [90] was demonstrated with the NASA Langley system by using the ruby laser wavelength of 6943Å to excite molecular constituents of the atmosphere and cause Raman scattering at other wave lengths associated with these molecules. Figure 7.10 illustrates the wavelengths excited for SO_2, NO, N_2, and H_2O.

Figure 7.11 shows the water vapor mixing ratio to 2.4 km altitude; here, the data points represent the ratio of Raman water vapor to Raman nitrogen backscatter signal.

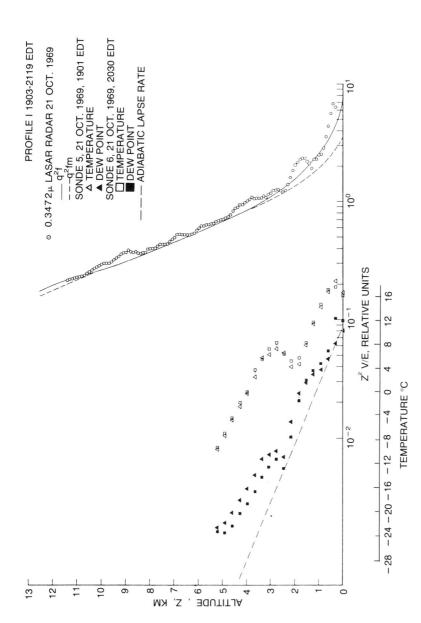

Figure 7.8 $Z^2 V/E$ *versus* Z laser radar data taken from 1903 to 2119 EDT on October 21, 1969, for $\lambda = 0.3472$ μm. The data are normalized to the predicted profiles. Also shown are the temperature and dew point temperature profiles from radiosondes launched at 1901 and 2030 EDT on October 21, 1969, and an adiabatic lapse rate curve.

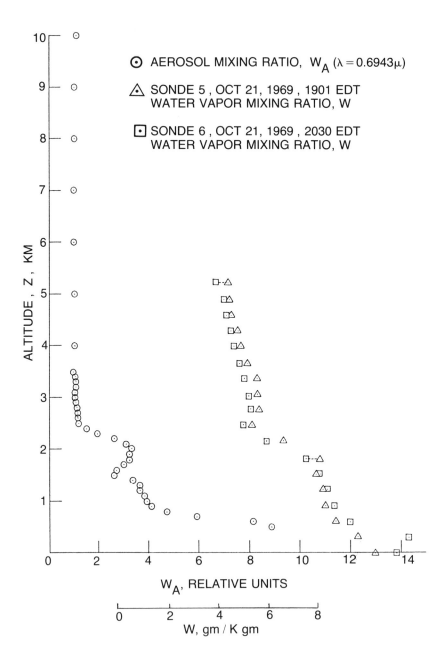

Figure 7.9 Comparison of aerosol mixing ratio for the $\lambda = 0.6943$ μm data of Figure 7.6 and water vapor mixing ratio profiles of sondes 5 and 6. The aerosol mixing ratio data are normalized to unity for its lowest value.

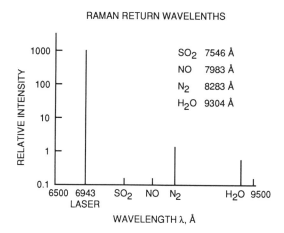

Figure 7.10 Stokes line positions of molecules in the atmosphere (in wavelength units excited at 6943Å).

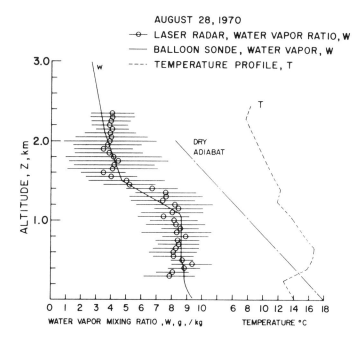

Figure 7.11 Comparison of water vapor mixing ratio measured with Raman Lidar and balloon-sonde.

7.2.3 Coherent Systems

Coherent detection systems use the temporal and spatial coherence of a laser radar (LIDAR) to obtain Doppler backscattering from aerosols and particulate matter normally suspended in the atmosphere to monitor the atmosphere. These systems operate both in the near field and far field of the system [91–102].

7.2.3.1 Near-Field Operation

In the near field, optical resolution is obtained by operating a coaxial telescope in a focused mode or by using bistatic apertures to illuminate and receive energy from a distinct volume of atmosphere.

Focused systems use the spatial coherence of the transceiver to obtain a resolved volume, which is determined by depth of field of the telescope, while bistatic systems rely on the common intersecting volumes of the transmitting and receiving beams. Sonnenschein and Horrigan [103] showed that the signal-to-noise ratio (SNR) of a focused system could be expressed as

$$\text{SNR} = \frac{\eta \, P_T B(\pi) \lambda \eta_{SYS} \eta_{ATM}}{2 \, Bhf} F(R,\lambda,D)$$

$$F(R,\lambda,D) = \left[\text{TAN}^{-1}\left(\frac{4\lambda R_2}{\pi D^2} - \frac{\pi D^2}{4\lambda R}\right)\left(1 - \frac{R_2}{R}\right) \right.$$
$$\left. - \text{TAN}^{-1}\left(\frac{4\lambda R_1}{\pi D^2} - \frac{\pi D^2}{4\lambda R}\right)\left(1 - \frac{R_1}{R}\right) \right] \quad (7.8)$$

where the term in parentheses represents the range-resolved region, and where R_2 and R_1 represent the region about the focus range R from which the return occurs. By defining the region between R_2 and R_1 as that region within which 50% of the total backscatter occurs in the near field, then

$$\text{SNR} = \frac{\pi P_T B(\pi) \lambda \eta_{SYS} \eta_{ATM}}{2 \, Bhf} \quad (7.9)$$

It should be noted that this expression is independent of the optics diameter and includes the diffuse target loss ($\eta_{SPECKLE}$), as the incoherent integration of random diffuse scatterers was included in the derivation.

The spatial resolution of a system having half the power obtained from the focused volume may be shown to be

$$\Delta R = \frac{8}{\pi} \frac{R^2 \lambda}{D^2} \qquad (7.10)$$

It may be observed that this spatial resolution is range-squared dependent, and as range increases the spatial extent quickly grows.

Assuming a 6-in aperture telescope ($D = 0.15$m), 100m spatial resolution (50% of return) could be obtained at 940 and 300m for system operating wavelength of 1.06 and 10.6 μm, respectively.

7.2.3.2 Far Field Operation

In the far field range, resolution may be obtained. Range resolution may also be obtained by using the temporal coherence of the system to provide range resolution by using a pulse transmitter where the range resolution is

$$\Delta R = \frac{C\tau}{2} \qquad (7.11)$$

Because of the Doppler effect, care must be taken in the choice of the pulse length (τ) since it also determines the velocity resolution of the system.

If the atmosphere is moving uniformly, the Doppler returns from the aerosols being transported by the wind field will provide a mean Doppler signal of

$$f_d = \frac{2V}{\lambda} \cos\theta \qquad (7.12)$$

which would provide a Doppler spectrum having a width determined by the atmosphere scatterers passing through the system beamwidth:

$$\Delta fd = \frac{2V}{\lambda} \sin\theta \, \Delta\theta \qquad (7.13)$$

where $\Delta\theta$ = transceiver beamwidth.

As the wind field becomes nonhomogeneous, turbulent, shear, *et cetera,* a differential velocity will occur over a range cell, which will provide a differential velocity to the Doppler return, and the Doppler spectrum will now include the effects of the velocity differential:

$$\Delta f_d = \frac{2V}{\lambda} \sin\theta \, \Delta\theta + \frac{2\Delta V}{\lambda} \cos\theta \qquad (7.14)$$

$$\left(\Delta f_d = B = \frac{1}{\tau}\right)$$

Matching the Doppler bandwidth Δf_d to the inverse of the pulse length ($1/\tau = B$) couples the range resolution, scattering volume, Doppler resolution, and scale of turbulence to be measured. This may be illustrated by

$$\textbf{Range resolution } \Delta R = \frac{C\tau}{2} \qquad (7.15)$$

$$\textbf{Velocity } \Delta V = \frac{\Delta f_d \lambda}{2} \quad (\theta = 0°)$$

or $\qquad (7.16)$

$$\Delta V = \frac{\lambda C}{4\Delta R}$$

Thus, as the range resolution increases (a longer atmospheric sample is measured), the velocity resolution decreases (smaller differential velocity is measured).

Sonnenschein and Horrigan [103] comprehensively evaluated the theoretical analysis related to the remote measurement of atmospheric aerosols and tabularized a variety of equations relating to operation in both the near and far field for systems that were focused or collimated.

The analysis used an untruncated aperture with a $1/e^2$ aperture illumination function. Table 7.1 illustrates the basic equations, where the

$$\text{SNR} = [\eta P_T \beta(\pi)\lambda]/(2Bh\nu) F(R,\lambda,f) \qquad (7.17)$$

and $F(R,\lambda,f)$ is the focal volume function found in Table 7.1 for pulsed and CW systems.

Substitution of the volumetric effective target cross section term into the far-field range equation will result in the equivalent SNR for an atmospheric return of

$$\text{SNR} = \frac{\pi^2 \, P_T B_{(\pi)} \, c\tau \, D^2 \, \eta_{SYS} \, \eta_{ATM}}{32 \, R^2 \, hf \, B} \qquad (7.18)$$

Table 7.1
SNR Variation (Courtesy of Applied Optics.)

System	$F(R, \lambda, f)$
Pulsed-Focused	$(\pi R_a^2 \Delta L)/(\lambda L_{AV}^2)$
Pulsed-Unfocused	$(\pi R_a^2 \Delta L)/\{\lambda L_{AV}^2[1 + (\pi R_a^2 \lambda L_{AV})^2]\}$
CW-Infinite Path-Focused	$\pi/2 + \tan^{-1}[(\pi R_a^2)/(\lambda f)]$
CW-Infinite Path-Unfocused	$\pi/2$
CW-Finite Path-Focused	$\tan^{-1}\left[\dfrac{\lambda L_2}{\pi R_a^2} - \dfrac{\pi R_a^2}{\lambda f}\left(1 - \dfrac{L_2}{f}\right)\right]$
	$-\tan^{-1}\left[\dfrac{\lambda L_2}{\pi R_a^2} - \dfrac{\pi R_a^2}{\lambda f}\left(1 - \dfrac{L_1}{f}\right)\right]$
CW-Finite Path-Unfocused	$\tan^{-1}[(\lambda L_2)/(\pi R_a^2)] - \tan^{-1}[(\lambda L_1)/(\pi R_a^2)]$
where R = aperture radius,	ΔL = Range-resolved distance
	L_{AV} = average range
$\Delta L = \Delta R$, $L_{AV} = R_{AVG}$	f = focal length

In this expression a diffuse (speckle) target loss of 3 dB (0.5 must be included in the system efficiency η_{SYS}) to be equivalent to those of Sonnenschein and Horrigan.

Assuming the atmospheric differential motion within a pulse does not result in bandwidth broadening, then the SNR expression may be shown to be dependent upon pulse length squared ($B = 1/\tau$).

Doppler broadening beyond that of a matched filter will result in an increase in the signal bandwidth and will therefore reduce the atmospheric signal in each filter from that of the unbroadened condition. The signal-to-noise ratio equation for this case will be linearly related to the pulse length. This Doppler broadening has been demonstrated to directly relate to atmospheric turbulence and the mean value to that of the atmospheric wind speed.

7.2.3.3 Near Field CW Doppler Experiments

Experimental field measurements suggest that a typical bandwidth of 100 kHz tends to match the normally quiescent atmosphere at a 10.6-μm wavelength. One such field test was directed toward the remote atmospheric measurement of aircraft trailing vortices.

Figure 7.12 illustrates a time sequence display of the velocity structure of an aircraft trailing vortex [96] obtained with a 20W laser and 15-cm optical system focused 100 ft into the atmosphere. The first picture shows the intensity of the received ground wind signal on a vertical scale, and the Doppler shift or velocity distribution of the ground wind signal on the horizontal scale. The succeeding pic-

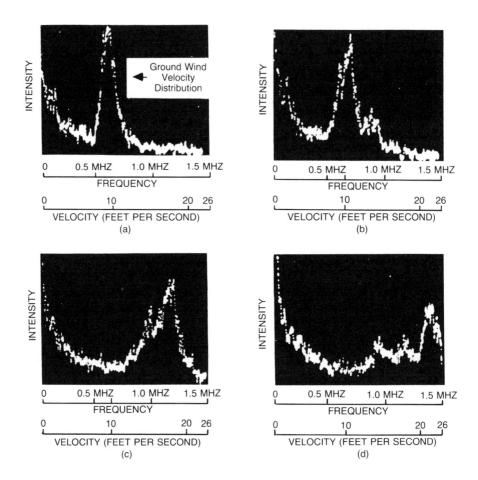

Figure 7.12 Time sequence display—aircraft trailing vortex, © IEEE. (Courtesy of *IEEE*.)

tures illustrate the shift in Doppler frequency and broadening of the spectrum as the wake vortex passes through the laser-sensitive volume. The last picture portrays the return to the normal ground wind spectrum. Figure 7.13 shows a photograph of a wake vortex entrained in smoke for visualization.

Backscatter measurements made at the 10.6-μm wavelength as a function of altitude have been reported by a variety of researchers and collected by Post of NOAA (Figure 7.14) [97]. These data illustrate the variability of the atmospheric scatter measurements when compared to the calculated theoretical models. M. Vaughan [102] and his colleagues at the Royal Signals and Radar Establishment in England extensively flight tested a coherent CO_2 laser radar, and Figure 7.15 illustrates a series of measurements as a function of altitude from their sensor.

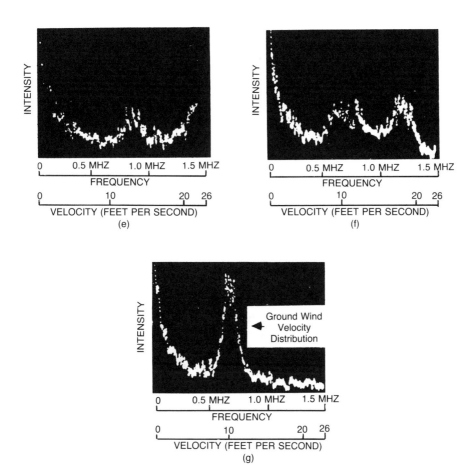

Figure 7.12 (cont'd)

Brandeiwie and Davis measured the atmospheric backscatter coefficient at 10.6 μm with a coherent bistatic laser radar [13], and their measurements are illustrated in Table 7.2.

7.2.3.4 Far-Field Pulse Doppler Measurements

Coherent far-field laser radar system operation may be illustrated by a block diagram of the airborne atmospheric pulse Doppler radar developed for NASA in 1970 [94,98–99] (Figure 7.16). Here a master oscillator power amplifier (MOPA) transmitter configuration is used to increase the coherent master oscillator power. Utilization of a

Figure 7.13 Wake vortex.

polarization-controlled optical modulator allows pulse waveform selection. The energy contained within a pulse is then propagated to the target, reradiated, and subsequently collected by the coaxial telescope, detected, and processed by a signal processor to provide range, velocity, and angle information. The unit was mounted to either look forward or through the side of a Convair 990 NASA aircraft in 1972.

Figure 7.17 illustrates the NASA 990 pod configuration to direct the beam forward of the aircraft. Table 7.3 illustrates some of the hardware specifications, while Figure 7.18 illustrates an A-scope (signal intensity as a function of range) display of a NASA Raytheon airborne pulse Doppler CO_2 laser radar having 10 kW of peak power. Here, in the lower part of the figure, there is an A-scope display, signal strength *versus* time where the x-axis is seen to have a range capability of 16 nmi. At approximately 13 nmi, an 8-dB SNR may be seen on this A-scope, illustrating that coherent CO_2 systems can be configured to operate in realistic environments and have sufficient coherence to perform long-range measurements.

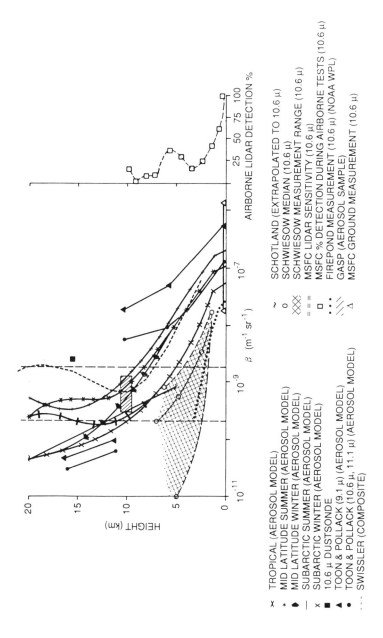

Figure 7.14 IR backscattering coefficient as a function of altitude.

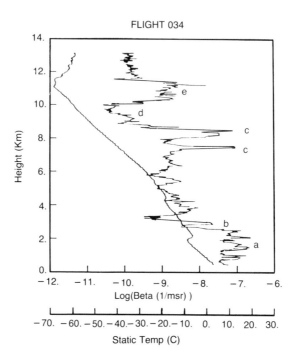

Figure 7.15 Backscatter and static air temperature *versus* altitude for a flight made in the Arctic Circle out of Bodo in Norway (~60°N, 16°E) on 19 March, 1986. (Courtesy of Applied Optics.)

Table 7.2
Measured Scattering Coefficients

Weather and Visibility	Scattering Coefficient
Weather to Heavy Overcast, Fog and Smog, Visibility 305m–610m)	4.121×10^{-7} m^{-1} ster^{-1}
Light Overcast, Visibility <8 km	2.36×10^{-7} m^{-1} ster^{-1}
Very Clear Air, Visibility >120 km	1.05×10^{-8} m^{-1} ster^{-1}

Figures 7.18–7.20 illustrate a velocity range (RVI) display where the velocity distributed in the atmosphere (ordinate) is portrayed as a function of the range in front of the aircraft (abscissa). In the lower half of the figure is an A-scope portrayal of the signal strength of the atmospheric return as a function of range. In Fig. 7.19 Run 20, the aircraft was in a runway approach and the atmospheric air speed is shown to be different than the ground speed at a range of 14 km. High-altitude returns (33,000 ft) are shown in Figure 7.20 to distances of greater than 10 km. In 1973, the sensor indicated a Doppler bandwidth broadening on an intensity-velocity (IVI)

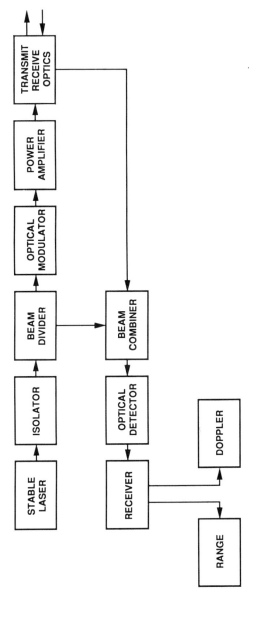

Figure 7.16 Block diagram of airborne pulse Doppler laser radar.

Figure 7.17 NASA Convair 990 aircraft in which pulse Doppler laser radar was mounted.

Table 7.3
NASA/Raytheon Pulse Doppler Laser Radar

System Parameters	
Operating wavelength	10.6 μm
Pulse width	Variable 1–10 μs
	Selectable pulses of 2, 4, and 8 μs
Repetition rate	200 pps
Optics size	12-in Cassegrain
Output polarization	Circular or vertical
Display	Range-velocity
Recording single range cell	Amplitude-velocity
Recording system	Analog or digital

(Courtesy of Raytheon Company, Sudbury, Mass., and NASA Marshal Space Flight Center, Huntsville, Ala.)

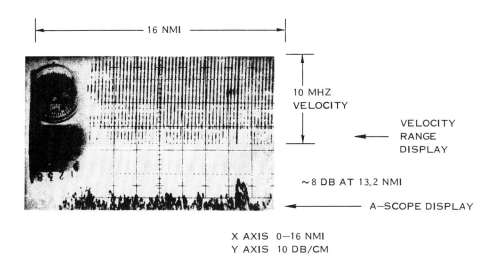

Figure 7.18 Pulse Doppler laser radar return. (Courtesy of Raytheon Co. and NASA Marshal Space Flight Center.)

CAT FLIGHT 21 - JANUARY 19, 1973

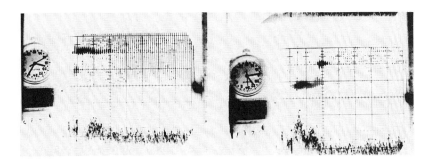

RUN 9 - IMPERIAL VALLEY

    DUST CLOUD AT 4,300 FT (1,311 METERS)
    HORIZ 2 nmi/DIV (3.71 km/DIV)
    VERT 10 dB/DIV A-SCOPE 2.64 MHz/DIV RVI

RUN 20 - OWENS VALLEY

    DUST CLOUD AND GROUND

Figure 7.19 Velocity range (RVI) display. (Courtesy of Raytheon Co. and NASA Marshal Space Flight Center.)

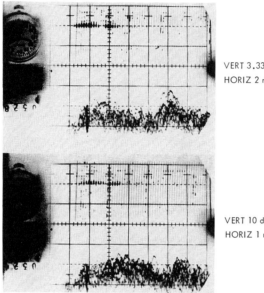

Figure 7.20 High-altitude cirrus returns. (Courtesy of Raytheon Co. and NASA Marshal Space Flight Center.)

display. Figure 7.21 illustrates the IVI display which samples the signal strength (ordinate) and velocity distribution (abscissa) at a range 2 miles in front of the aircraft. Bandwidth of the Doppler spectrum was broadened to 1 MHz, which was considered representative of a turbulent atmosphere of 5 m/s wind change over a 300m distance. The on-board accelerometer response (lower half of the figure) indicated between a 0.5 and a 1g response 22 seconds later when the aircraft flew into the general area measured by the laser radar.

When configured as a heterodyne system, that is, one laser for transmission, and a second automatic frequency-controlled laser supplying local oscillator power for the detector, the system was used to provide ground-based atmospheric wind shear information. This is shown in Figure 7.22, which illustrates a display of velocity distribution as a function of altitude. A model of a thunderstorm (Figure 7.22)

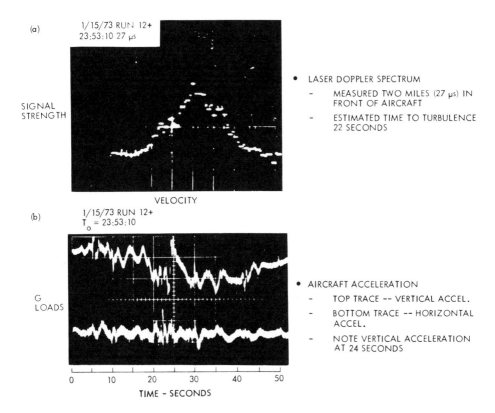

Figure 7.21 Intensity velocity (IVI) display. (Courtesy of Raytheon Co. and NASA Marshal Space Flight Center.)

is illustrated where a shaft of descending air approaches the ground and begins to spread, shown by the "outflow" arrows. The sea breeze is seen to be entering from the right to result in an updraft. Two locations of a laser Doppler velocimeter (LDV) van are shown relative to this flow field, even though only one system is used. This is done to illustrate the effect of spatially sampling this front. For a beam looking upward from location I, the flow field is initially a tail wind caused by the sea breeze, followed by a head wind and a tail wind as a function of altitude. Sampling of this thunderstorm from location II would result in a head wind caused by the outflow near the ground, followed by the updraft at altitude.

Measurements performed by an upward oriented conical scan pulsed laser system, sampling the atmosphere as a function of altitude and azimuth position (velocity azimuth display VAD) allow the wind strength and direction to be measured at altitude. The gust front measurements for such a system are shown in the lower right (Figure 7.22), as observed from location I.

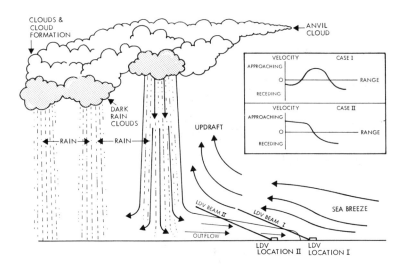

Figure 7.22 Wind shear measurements display. (Courtesy of Raytheon Co. and NASA Marshal Space Flight Center.)

References

[1] D. Klein, "Optical Antenna Gain. 1: Transmitting Antennas," *Appl. Opt.*, Vol. 13, September 1974.
[2] W.F. Wolfe, *Handbook of Military IR Technology*, Washington, D.C.: U.S. Government Printing Office, 1965.
[3] M.W.P. Standberg, "Inherent Noise of Quantum Mechanical Amplifiers," *Physics Review*, Vol. 106, May 15 1957, pp. 610–617.
[4] *Electro-Optics Handbook*, RCA Commercial Engineering, Harrison, N.J., May 1968.
[5] K. Seeber, Raytheon Company, Sudbury, Mass. December 1987 (private correspondence).
[6] D.K. Barton, "Low Angle Radar Tracking," *Proc. IEEE*, Vol. 62, No. 6, June 1974.
[7] P.E. Cornwell, "Multipath Measurements at 140 GHz," Report No. PEC 79:71, Raytheon Company, Wayland, Mass.
[8] B. Wallace, "140 GHz Multipath Measurement Over Varied Ground Clutter," *EASCON 79*.
[9] D.T. Hayes, *et al.*, "Millimeter Wave Propagation Measurements Over Sayon," *EASCON 79*.
[10] J.C. Leader, "Speckle Effects on Coherent Laser Radar Detection Efficiency," *Opt. Eng.*, May 25, 1986.
[11] D. Gabor, "Laser Speckle and Its Elimination," *IBM Journal Research and Development*, September 1970, pp. 509–514.
[12] J.M. Cruickshank, *et al.*, "Field Measurements with a Coherent Transversely Excited Atmospheric CO_2 Laser Radar," *SPIE*, Vol. 415, April 7–8, 1983.
[13] R.A. Brandeiwie, and W.C. Davis, *Parametric Study of a 10.6 μm Laser Radar Applied Optics*, Vol. II, No. 7, July 1972, pp. 1520–1533.
[14] J.H. Shapiro, B.A. Capron, and R.C. Harney, "Imaging and Target Detection with a Heterodyne-Reception Optical Radar," *Appl. Opt.*, Vol. 20, Oct. 1981, p. 3292.
[15] W.L. Wolfe, and G.J. Zissis, *The Infrared Handbook*, Washington, D.C.: U.S. Government Printing Office, 1979.
[16] M. Ross, *Laser Receivers*, New York: John Wiley & Sons, 1966.
[17] Forrestor, *Physics Review 90*, 1955, p. 1691.
[18] W.S. Beard, and D.L. Fried, "Optics Heterodyning with Non-Critical Angular Alignment," *Proc. IEEE*, Vol. 51, December 1963, p. 1781.
[19] K. Seeber, Raytheon Company, (private communication).
[20] J.J. Degnan, and B.J. Klein, "Optical Antenna Gain. 2: Receiving Antennas," *Appl. Opt.*, Vol. 13, October 1974.
[21] S.C. Cohen, "Heterodyne Detection: Phase Front Alignment, Beam Spot Size, and Detector Uniformity," *Appl. Opt.*, Vol. 14, August 1975.

[22] J.H. Shapiro, "Precise Comparison of Experimental and Theoretical SNR's in CO_2 Laser Heterodyne Systems: Comments," *Appl. Opt.,* Vol. 24, No. 9, May 1, 1985.

[23] R.S. Lawrence, G.R. Osche, and S.F. Clifford, "Measurements of Atmospheric Turbulence Relevant to Optical Propagation," *Journal of the Optical Society of America,* Vol. 66, June 1970, pp. 826–830.

[24] E. Brookner, "Atmospheric Propagation and Communication Channel Model for Laser Wavelengths—an Update," IEEE Transactions on Communication Technology, Vol. COM-22, February 1974, pp. 265–270; and E. Brookner, "Atmospheric Propagation and Communication Channel Model for Laser Wavelengths," *IEEE Transactions on Communication Technology,* Vol. COM-18, August 1970, pp. 396–416. Both articles are reprinted in B. Goldberg, ed., *Communication Channels: Characterization and Behavior,* IEEE Press, 1976.

[25] R.E. Huffnagel, "An Improved Model Turbulent Atmosphere in Restoration of Atmospherically Degraded Images," Vol. 2 of *Woods Hole Summer Study,* Appendix 3, Woods Hole, Mass., July 1966, pp. 15–18.

[26] D.L. Fried, "Optical Heterodyne Detection of an Atmospherically Distorted Signal Wave Front, *Proc. IEEE,* Vol. 55, January 1967, pp. 57–66.

[27] I. Goldstein, P.A. Miles, and A. Chabot, "Heterodyne Measurements of Light Propagation Through Atmospheric Turbulence," *Proc. IEEE,* Vol. 53, September 1965, pp. 1172–1180.

[28] C. Freed, "Ultrastable CO_2 Lasers," *The Lincoln Laboratory Journal,* Vol. 3, No. 3, Fall 1990, pp. 479–499.

[29] J.J. Zayhowski, "Microchip Lasers," *The Lincoln Laboratory Journal,* Vol. 3, No. 3, Fall 1990, pp. 427–445.

[30] M.J. Skolnik, "Theoretical Accuracy of Radar Measurements," *IRE Transactions on Aeronautical and Navigational Electronics.*

[31] C.E. Cook, "Pulse Compression—Key to More Efficient Radar Transmissions," *Proc. IRE,* March 1960, pp. 310–316.

[32] M. Elbaum, and M. Greenebaum, "Annual Apertures for Angular Tracking," *Appl. Opt.,* Vol. 16, September 1977.

[33] D. Youmans, "Propagation to Space," 1988 (private communication).

[34] E. Gatti, and S. Donnati, "Optimum Signal Processing for Distance Measurement with Lasers," *Appl. Opt.,* Vol. 10, November 1971.

[35] S. Weiner, MIT Lincoln Laboratory, Lexington, Mass., June 1986, (private correspondence).

[36] A.L. Kachelmyer, "Range-Doppler Imaging with a Laser Radar," *The Lincoln Laboratory Journal,* Vol. 3, No.1, Spring 1990, pp. 87–118

[37] Kachelmyer, A.L. "Range-Doppler Imaging: Waveforms and Receiver Design," *SPIE,* Vol. 999, 1988, p. 138.

[38] A.L. Kachelmyer, R.E. Knowlden, and W.E. Keicher, "Effect of Atmospheric Distortion of Carbon Dioxide Laser Radar Waveforms," *SPIE,* Vol. 783, 1987, p. 101.

[39] A.E. Siegman, "The Antenna Properties of Optical Heterodyne Receivers," *Appl. Opt.* Vol. 5, 1966, p. 1588.

[40] R.M. Gagliardi, and S. Karp, "Optical Communications," New York: John Wiley & Sons, 1976.

[41] J.H. Shapiro, "The Target-Reflectivity Theory for Coherent Laser Radars," *Appl. Opt.,* Vol. 21, 1982, p. 3398.

[42] F.K. Knight, S.R. Kulkarni, R.M. Marino, and J.K. Parker, "Tomographic Techniques Applied to Laser Radar Reflective Measurements," *The Lincoln Laboratory Journal,* Vol. 2, 1989, p. 143.

[43] Santa Barbara Research Center wallchart.

[44] R. McClatchey, J. Selby, "Environmental Research Paper No. 419," AFCRL-72-0611, AFCRL, Bedford, Mass., 12 October 1972.

[45] F. Goodwin and T. Nussmeier, *IEEE Journal of Quantum Electronics*, Vol. QE-4, No. 10, October 1968, p. 616.
[46] Rensch and Long, *Applied Optics*, Vol. 9, No. 7, July 1970, pp. 1563–1573.
[47] T.S. Chu, "Bell System Technical Journal," May/June 1968, pp. 723–759.
[48] D.K. Barton, *Radar System Analysis*, Dedham, Mass.: Artech House, 1976.
[49] M.I. Skolnik, *Introduction to Radar Systems*, New York: McGraw-Hill, 1962.
[50] R.A. McClatchey, and J.E.A. Selby, "Atmospheric Attenuation of Laser Radiation From 0.76 to 31.25 μm," AF Research Laboratory Report TR-74-0003, Bedford, Mass., January 1974.
[51] L.A. Young, "Infrared Spectra," *Journal of Quantitative Spectroscopy and Radiative Transfer*, Oxford: Pergamon Press, Vol. 8, No. 2, February 1968, pp. 693–716.
[52] A.W. Mantz, E.R. Nichols, D.B. Alpert, and K.N. Rao, "CO_2 Laser Spectra Studied with a 10-Meter Vacuum Infrared Grating Spectrogram," *Journal of Molecular Spectroscopy*, New York: Academic Press, Vol. 35, 1970, p. 325.
[53] T.F. Deutsch, "Molecular Laser Action in Hydrogen and Deuterium Halides," *Applied Physics Letters*, New York: American Institute of Physics, Vol. 10, 1971, p. 234.
[54] N.G. Basov, V.T. Galochkin, V.I. Igoshin, L.V. Kulakov, E.P. Martin, A.J. Nitikin, and A.N. Oraevsky, "Spectra of Stimulated Emission in the Hydrogen-Fluorine Reaction Process and Energy Transfer from DF to CO_2," *Appl. Opt.*, Washington, D.C.: Optical Society of America, Vol. 10, 1971, p. 1814.
[55] R.N. Spanbauer, K.N. Rao, L.H. Jones, *Journal of Molecular Spectroscopy*, New York: Academic Press, Vol. 16, No. 1, May 1965, p. 100.
[56] J.H. McCoy, "Atmospheric Absorption of Carbon Dioxide Laser Radiation Near Ten Microns," The Ohio State University, Electro-Science Laboratory, Columbus, Ohio, Contract F33(615)-67-C-1949, Air Force Avionics Lab, Air Force Systems Command, U.S. Air Force, 1968.
[57] A. Jelalian, W. Keene, D. Kawachi, C. Sonnenschein, N. Freedman, and C. DiMarzio, "Low Probability of Intercept Multifunctional Tactical Sensors," Volume 2, Phase II, AFWAL-TR-80-1006, Raytheon Company, Sudbury, Mass., March 1979.
[58] C.C. Chen, "A Correction for Middleton's Visibility and Infrared-Radiation Extinction Coefficient Due to Rain," Rand Corp. R-1523-PR., 1974.
[59] T.S. Chu, and D.C. Hogg, "Effects of Precipitation of Propagation at 0.63, 3.5 and 10.6 Microns," BSTJ47, No. 5, 1968.
[60] C.A. DiMarzio, R.W. Goff, A.V. Jelalian, and R.M. Kalafus, "Coherent CO_2 Lidar Attenuation Measurements in Rain," presented at the IEEE/OSA Topical Meeting on Optical Propagation Through Turbulence, Rain and Fog, University of Colorado, Boulder, Colo., August 9–11, 1977.
[61] D.G. Rensch, and R.K. Long, "Comparative Studies of Extinction and Backscattering by Aerosols, Fog and Rain at 10.6 Microns and 0.63 Microns," *Appl. Opt.*, Vol. 9, No. 7.
[62] V.P. Bisyarin, I.P. Bisyarina, V.K. Rubash, and A.V. Sokolov, "Attenuation of 10.6 and 0.63 Micron Laser Radiation," *Radio Engineering and Electronic Physics*, Vol. 16, Oct. 1971, pp. 1594–1597.
[63] K. Seeber, and G. Osche, "The Mathematical Distinctions Between Microwave Radar and Laser Radar Detection Statistics," 1989 (private communication).
[64] J.I. Marcum, *A Statistical Theory of Target Detection by Pulsed Radar: Mathematical Appendix*, the RAND Corporation, Research Memorandum RM-753, July 1, 1948.
[65] P. Swerling, "Probability of Detection of Fluctuating Targets," RAND Corporation Report RM-1217, March 17, 1954.
[66] P. Swerling, "Probability of Detection for Fluctuating Targets," *IRE Trans.*, Vol. IT-6, April 1960, pp. 269–308.
[67] W.M. Hall, "General Radar Equation," Raytheon Report.

[68] B.A. Capron, R.C. Harney, J.H. Shaprio, "Turbulence Effects on the Receiver Operating Characteristics of a Heterodyne Reception Optical Radar," Project Report TST-33, MIT Lincoln Laboratory, 1979.
[69] V.I. Tatarski, "Wave Propagation in a Turbulent Medium," R.A. Silverman, trans., New York: McGraw-Hill, 1961.
[70] Fante, "Propagation in Turbulent Media," *Proc. IEEE,* December 1975.
[71] R.L. Paport, J.H. Shapiro, R.C. Harney, "Physics and Technology of Coherent Infrared Radar," *SPIE,* Vol. 30, 1981.
[72] J.W. Goodman, "Some Effects on Target-Induced Scintillation on Optical Radar Performance," *Proc. IEEE,* Vol. 53, No. 11, Nov. 1965, pp. 1688–1700.
[73] J. Goodman, "Comparative Performance of Optical-Radar Detection Techniques," *IEEE Transactions on Aerospace and Electronic Systems,* Vol. AES-2, No. 5, September 1962.
[74] K.N. Seeber, C.A. DiMarzio, "Incoherent Optical Signal Detection in Gaussian Noise," (private communication).
[75] *The Lincoln Laboratory Journal,* Vol. 3, No. 3, Fall 1990.
[76] ITT wall chart.
[77] "Laser Focus Buyer's Guide," 1977.
[78] J.H. Woodward, "The AN/GVS-5 Hand-Held Laser Rangefinder," RCA Electro-Optics, Moorestown, N.J., 1977, p. 76–79.
[79] G.R. Osche, D.S. Young, W.J. Wilson, STARTLE Technology Demonstrator System, Raytheon Company, Sudbury, Mass., IR DAAK70-80-C-0192, August 19, 1986.
[80] A.V. Jelalian, D. Kawachi, and C. Miller, "Radar Electronic Counter-Countermeasures," *Electro 1976,* Boston, Mass., 1976.
[81] T. Gilmartin, H. Bostick, and L. Sullivan, Nerem Record, MIT Lincoln Laboratory, Lexington, Mass., May 1973.
[82] R. Kingston, and L. Sullivan, "Optical Design Problems in Laser Systems," *SPIE,* Vol. 69, 1975.
[83] R. Teoste, and W.J. Scouler, MIT Report 1976-19A, "10.6 μm Coherent Monopulse Tracking," Lincoln Laboratory, Lexington, Mass., May 7, 1976.
[84] W. Keicher, SDIO Firefly Experiment, Briefing, April 10, 1990.
[85] Aviation Week and Space Technology, April 23, 1990, p. 75.
[86] J.D. Lawrence, M.P. McCormick, S.N. Melfi, and P.P. Woodman, "Optical Studies of the Atmosphere," Fifth Symposium on Remote Sensing of the Environment, Ann Arbor, Mich., April 16–18, 1968.
[87] M.P. McCormick, "Simultaneous Multiple Wavelength Laser Radar Measurements of the Lower Atmosphere," Electro-Optics International Conference, Brighton, England, March 25, 1971.
[88] M.L. Wright, E.K. Proctor, L.S. Gasiorek, and E.M. Liston, "A Preliminary Study of Air Pollution Measurements by Active Remote Sensing Techniques," NASA CR-132724, 1975.
[89] R.M. Measures, *Laser Remote Sensing,* New York: John Wiley & Sons, 1984.
[90] S.H. Melfi, "Remote Measurements of the Atmosphere Using Raman Scattering," *Appl. Opt.,* Vol. 11, 1972, pp. 1605–1610.
[91] W.A. McGowan, "Remote Detection of Turbulence in Clear Air," AIII Air Transportation Conference, Key Biscayne, Fla., June 8, 1971.
[92] E.A. Weaver, "Clear Air Turbulence Using Lasers," NASA SP-270, May 1970.
[93] R.M. Huffaker, "Laser Doppler Detection for Gas Velocity Measurements," *Appl. Opt.,* Vol. 9, No. 5, May 1970, pp. 1026–1039.
[94] A.V. Jelalian, W.H. Keene, and C.M. Sonnenschein, "Development of CO_2 Laser Doppler Instrumentation for the Detection of Clear Air Turbulence," Final Report NAS 8-24742, Raytheon Company Report ER-70-4203, June 5, 1970.

[95] A.V. Jelalian, and R.M. Huffaker, "Specialist Conference on Molecular Radiation and its Applications to Diagnostic Techniques," NASA T.M. X-53711, October 5–6, 1967, p. 345.

[96] R.M. Huffaker, A.V. Jelalian, and A.F. Thomson, "Laser Doppler System for Detection of Aircraft Trailing Vortices," *Proc. IEEE*, Vol. 58, No. 3, March 1970.

[97] J.W. Bilbro, "Atmospheric Laser Doppler Velocimetry: An Overview," *Opt. Engr.*, Vol. 19, July and August 1980, pp. 533–542.

[98] J.W. Bilbro, C.A. Dimarzio, D. Fitzjarrald, S. Johnson, and W. Jones, "Airborne Doppler Lidar Measurements," *Appl. Opt.*, Vol. 25, No. 21, November 1986, pp. 3952–3960.

[99] C.A. Dimarzio, C.E. Harris, J.W. Bilbro, E.A. Weaver, D.C. Burnham, and J.N. Hallock, "Pulsed Laser Doppler Measurements of Wind Shear," Bulletin of American Meteorological Society, September 1979, pp. 1061–1065.

[100] H.B. Jeffreys, J.W. Bilbro, C.A. Dimarzio, C.M. Sonnenschein, and D.W. Toomey, "Laser Doppler Vortex Measurements at John F. Kennedy International Airport," presented at the 17th Conference on Radar Meteorology, American Meteorological Society, Seattle, Wash., October 1976.

[101] H.W. Mocker, and P.E. Bjork, "High Accuracy Laser Doppler Velocimeter Using Stable Long-Wavelength Semiconductor Lasers," *Appl. Opt.*, Vol. 28, No. 22, November 1989, pp. 4914–4919.

[102] J.M. Vaughan, R.P. Callan, D.A. Bowdle, and J. Ruthermel, "Spectral Analysis, Digital Integration and Measurement of Low Backscatter in Coherent Laser Radar," *Appl. Opt.*, Vol. 28, No. 15, August 1, 1989.

[103] C.M. Sonnenschein, and F.A. Horrigan, "Signal to Noise Relationships for Coaxial Systems that Heterodyne Backscatter from the Atmosphere," *Appl. Opt.*, Vol. 10, No. 7, July 1971, pp. 1600–1604.

[104] M. Teich, and B. Saleh, "Effects of Random Detection and Adaptive Noise on Bunched and Antibunched Photon Counting Statistics," Opt. Lett, Vol. 7, No. 9, Aug. 1982.

*Appendix A**

A.1 INTRODUCTION

Target Reflection Characteristics

Laser radar systems design requires a balancing of sensor parameters such as power, wavelength, aperture size, *et cetera*, in order to provide optimal sensor design. Key to this optimization is an appreciation of the target signature. In the 1960s the United States Air Force Avionics Laboratory at Wright-Patterson Air Force Base, Dayton, Ohio, contracted with the University of Michigan Willow Run Laboratories at Ann Arbor to establish a target signature analysis center for providing a data compilation of reflectance data covering the visible, near- and far-IR spectrum. This data compilation was published in 1966 and 1967 and was authored by Dianne George Earling and James Smith from the University of Michigan.

A discussion of the reflectance theory and selected data from this report is included here for reference purposes. The materials selected include metals, paints, trees/leaves, soils, and building materials.

- *Metals*:
  Aluminum
  Stainless Steel
  Titanium
  Chromium
  Iron
  Iconel
  Brass/copper
  Gold/substrate
  Magnesium
  Nickel

---

*This appendix is reproduced with the permission of the USAF Wright Research and Development Center and the Environmental Research Institute of Michigan.

    Nickel aluminum
    Platinum aluminum
    Carbon black
- *Paint:*
    Black
    Blue
    Enamel
- *Carbonate/nitrates/chlorides:*
    Calcium
    Sodium
- *Trees/leaves:*
    Maple
    Asphalt
    Building materials
    Coal tar
    Mud/soil
    Loam
    Water

A.2 DISCUSSION OF REFLECTANCE MEASUREMENTS

A.2.1 Theory

The purpose of this section is to enable the user of this data compilation to consider the data in a proper perspective. The reflectance alone, for example, does not sufficiently describe the results of an experiment, as will become obvious in this section. It is necessary to have knowledge of the measuring instrument's characteristics, since they have measurable effect on interpretation of the output. Some important instrument parameters include spectral resolution, the solid angle of effective viewing, and characteristics of the radiation source.

Our present understanding of radiation theory does not permit an analytical description, in closed form, of the exact relationship between the radiation emitted by a source (whether natural or artificial) and the radiation received by a remote sensor after having been reflected by an object under surveillance. There are well known laws to describe the simple case of an electromagnetic wave incident upon a perfectly planar interface between two media. In this case, the reflected wave depends upon the radiation wavelength, the angle of incidence, and the physical properties (permittivity, permeability, and conductivity) of the two adjoining media. The laws governing such a case are sufficiently understood so that the refractive index and extinction coefficient of materials involved may be found by determining the reflection coefficients of the materials. For the more complicated case involving a surface with periodic or random surface irregularities, an analytic determination of the properties of the reflected electromagnetic field may only be approximated.

In the past ten years, many papers have been published on scattering, or reflection from rough surfaces. Many theories have been developed, but none is both general and rigorous at the same time. To perform reasonably simple numerical calculations on the basis of these theories, certain simplifying assumptions are introduced, usually including one or more of the following:

- The dimensions of scattering elements of the rough surface are either much smaller or much greater than the wavelength of the incident radiation.
- The radii of curvature of the scattering elements are much greater than the wavelength of the incident radiation.
- Shadowing or obscuration effects occurring at the surface may be neglected.
- Only the far field is to be considered.
- Multiple reflections may be neglected.
- Consideration is restricted to a particular model of surface roughness (e.g., sawtooth, sinusoidal protrusions of definite shape and in random position, with random variations in height given by their statistical distribution and correlation function).

Electromagnetic scattering theory has been used in the past to compute radiation backscattering from targets in the microwave region of the spectrum, where the radiation wavelength is much greater than the minute irregularities of the target surface, and where the conductivity of the target material is infinite. In the optical region, where materials have finite conductivity and the surface irregularities have a wide range in size relative to the radiation wavelength, present electromagnetic scattering theory is applicable to only a few special cases, so the only way to determine reflectance in this region for target and background objects is by experimentation.

It is possible to arrive at the most general definition of reflectance ρ' (called bidirectional reflectance) [1]* by considering an infinitesimal element of surface, dA, upon which radiation of infinitesimal solid angle $d\omega_i$ and radiance N_i are incident. Taking a coordinate system fixed with respect to dA, with polar angle θ' measured from the normal and azimuth angle ϕ', measured from a fixed line (see Figure A.1), the contribution to the reflected radiance, $dN_r(\theta'_r, \phi'_r)$, in the reflected pencil for the direction (θ'_r, ϕ'_r) is

$$dN_r(\theta'_r, \phi'_r) = \rho' N_i(\theta'_i, \phi'_i) \cos\theta'_i d\omega_i \quad (A.1)$$

Generally, ρ' is a function of the incident and reflected directions (θ'_i, ϕ'_i and θ'_r, ϕ'_r, respectively), the polarization (P), the wavelength (λ), and the optical parameters of

---

*The definitions presented in this report conform to those proposed by D. Klein [1].

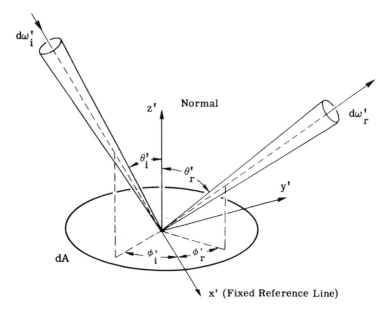

Figure A.1 Local coordinate system for determining bidirectional reflectance.

the material on either side of the surface. Total radiance in a given reflected direction is obtained by integrating Equation (A.1) over all incident directions, which yields

$$N_r(\theta'_r, \phi'_r) = \int \rho' N_i(\theta'_i, \phi'_i) \cos\theta'_i \, d\omega'_i \qquad (A.2)$$

Also, by Helmholtz's reciprocity theorem, if the directions of the incident and reflected pencils are interchanged, the bidirectional reflectance is unchanged:

$$\rho'(\theta'_1, \phi'_1; \theta'_2, \phi'_2; P; \lambda) = \rho'(\theta'_2, \phi'_2; \theta'_1, \phi'_1; P; \lambda) \qquad (A.3)$$

Since the optical constants of materials may change from point to point, bidirectional reflectance becomes a function of the location of dA. If it is then assumed that the surface can be described by the $z' = f(x', y')$, the correct functional dependence for reflectance is

$$\rho'(\theta'_i, \phi'_i; \theta'_r, \phi'_r; P; \lambda; x', y', z')_{z'=f(x',y')} \qquad (A.4)$$

Generally, the direction of the normal to dA is also a function of the location of dA on the surface of the object. Hence, even if the incident and reflected radiation have a constant direction with respect to the (x', y', z') coordinates, the angles (θ'_i, ϕ'_i) and (θ'_r, ϕ'_r) (taken with respect to the local normal) would be a function of location of the surface element dA. For convenience, a second, absolute coordinate system is usually introduced, namely, (x, y, z). The x-y plane of this system is coincident with the average value of $z' = f(x', y')$ along the surface A, and is, therefore, the average plane of the reflector. The normal to this average plane is parallel to the z axis. Instead of referring the incident and reflected radiation to the local coordinates, they are then referred to the absolute system, with θ as the polar angle and ϕ as the azimuthal angle. The bidirectional reflectance with respect to this system is

$$\rho'(\theta_i, \phi_i; \theta_r, \phi_r; P; \lambda; x, y) \qquad (A.5)$$

Another type of reflectance commonly considered is the directional reflectance ρ_d, which is a function of only one direction, either the incident or reflected direction. In the case where reflected power is integrated over a hemisphere and incident power is from a specific direction, directional reflectance is denoted by ρ_{di}. The incident power dp_i is

$$dp_i = dN_i(\theta_i, \phi_i; P_i) \cos\theta_i \, d\omega_i \, dA \qquad (A.6)$$

and, using Equation (A.2),

$$dN_r = \rho' \frac{dp_i}{dA} \qquad (A.7)$$

Since the reflected power dp_r is given by

$$dp_r = dA \int_{2\pi} dN_r \cos\theta_r d\omega_r = dp_i \int_{2\pi} \rho' \cos\theta_r d\omega_r \qquad (A.8)$$

therefore,

$$\rho_{di}(\theta_i, \phi_i; P; \lambda; x, y) = \int_{2\pi} \rho' \cos\theta_r \, d\omega_r \qquad (A.9)$$

When dA is uniformly illuminated from all directions (N_i = constant), the corresponding directional reflectance, ρ_{dr}, is defined as the ratio of the radiance reflected in a given direction to the incident radiance. To proceed as previously,

$$N_r = \int_{2\pi} \rho' N_i \cos\theta_i \, d\omega_i = N_i \int_{2\pi} \rho' \cos\theta_i d\omega_i \qquad (A.10)$$

and, thus,

$$\rho_{dr}(\theta_r, \phi_r; P; \lambda; x,y) = \int_{2\pi} \rho' \cos\theta_i \, d\omega_i \tag{A.11}$$

From comparison of Equations (A.8) and (A.9),

$$\rho_{di}(\theta, \phi; P; \lambda; x,y) = \rho_{dr}(\theta, \phi; P; \lambda; x,y) = \rho_d \tag{A.12}$$

ρ_d is called total directional reflectance.

Metals

220

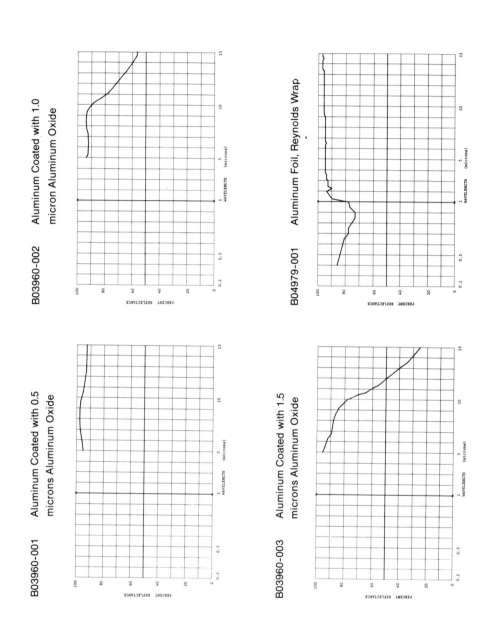

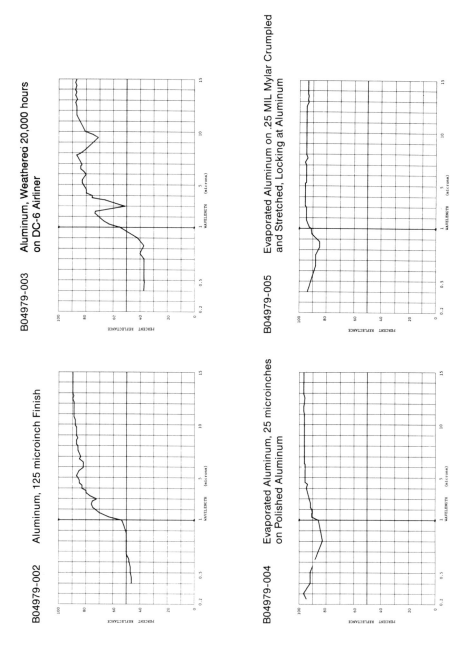

B04979-007 Aluminum, Commercially Pure, 300 hours at 600°F

B04979-008 Aluminum, Commercially Pure

B04979-009 Aluminum, Commercially Pure, 300 hours at 1000°F

B04979-010 Chromic Acid Anodized on Aluminum

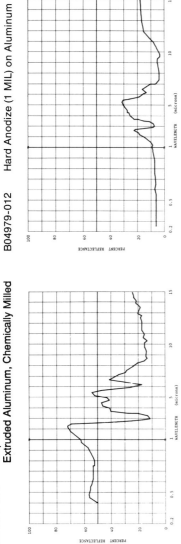
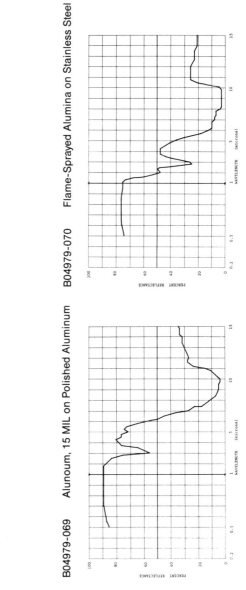

B04979-121 Aluminum Foil

B04979-122 Oxided Evaporated Titanium on Aluminum Foil, Oxidized at 400°C

B04979-123 Anodized Aluminum, Sulfuric Acid Anodize

B04979-124 Flame-Sprayed Alumina on 410 Stainless Steel

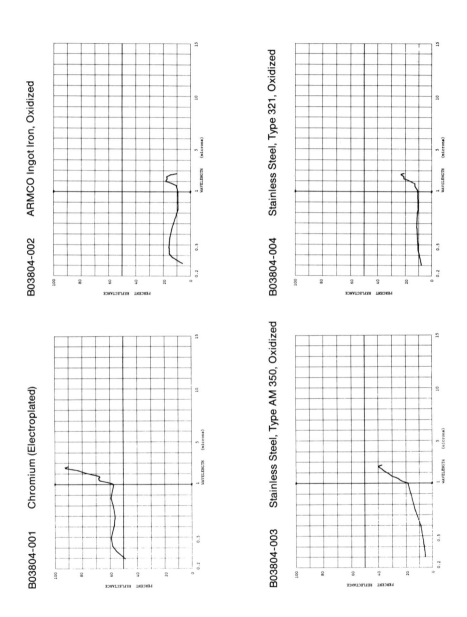

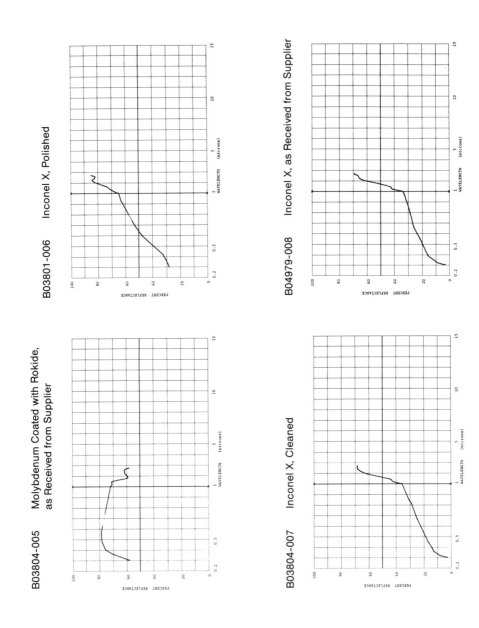

227

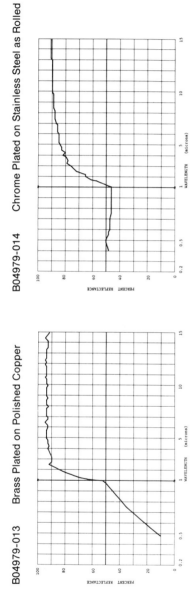

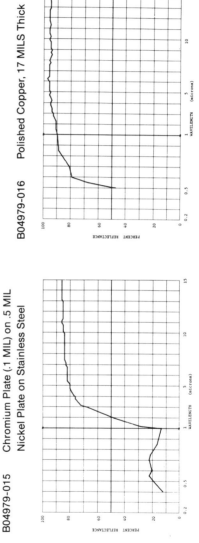

228

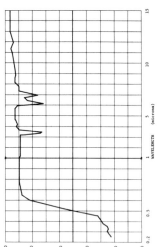

B04979-017 Immersion Gold Approx. .03 MIL on Nickel Plated Copper Plate on Polished Aluminum

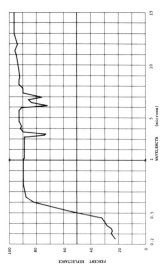

B04979-018 Vacuum Evaporated Gold on Fiberglass Laminate

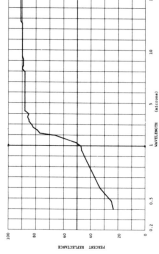

B04979-020 White Gold on Polished Steel

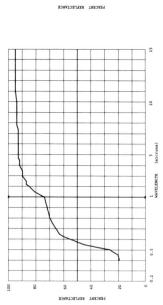

B04979-019 Gold Ash (80 microns) on .4 MIL Silver on Epon Glass

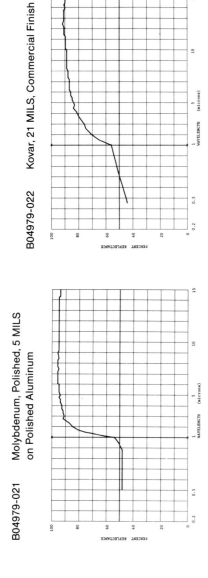
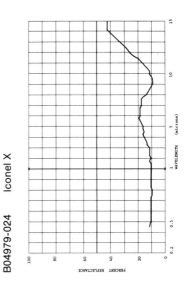
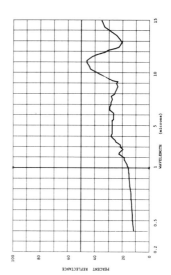

230

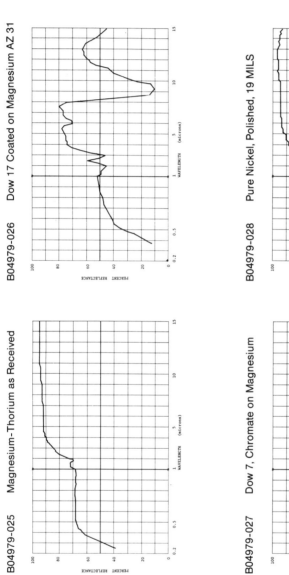
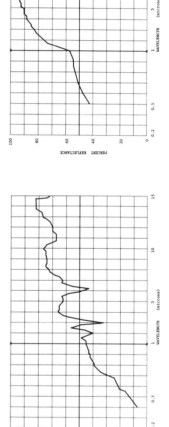

231

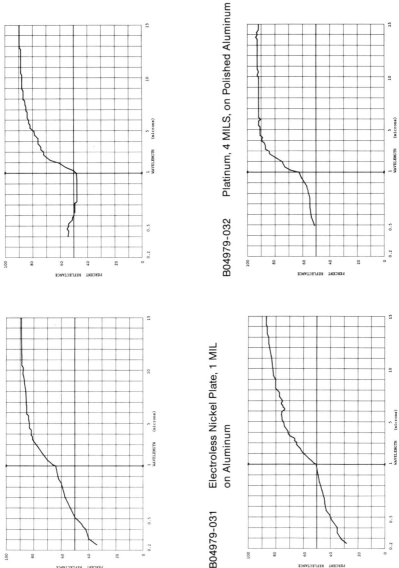

232

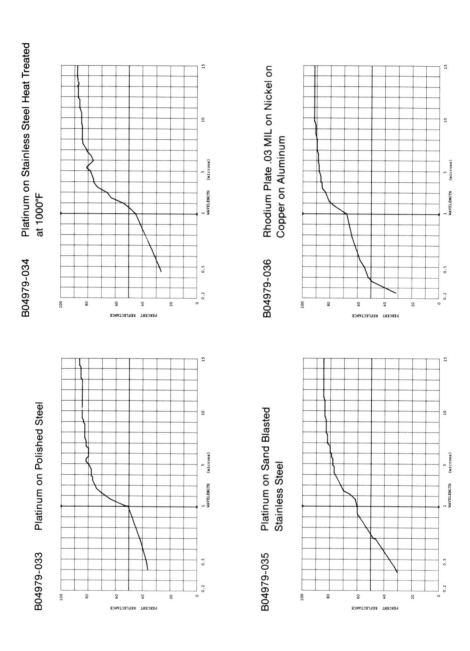

233

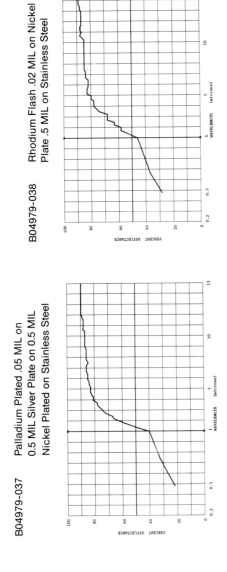
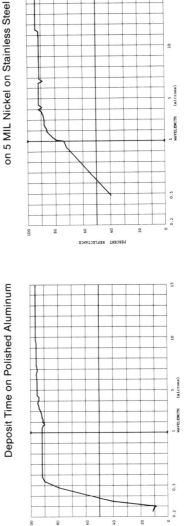

234

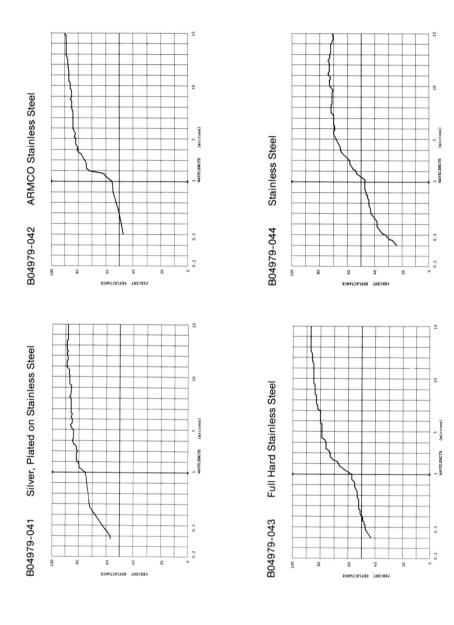

235

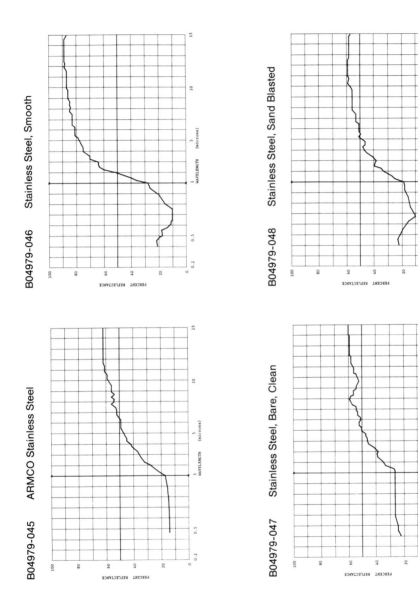

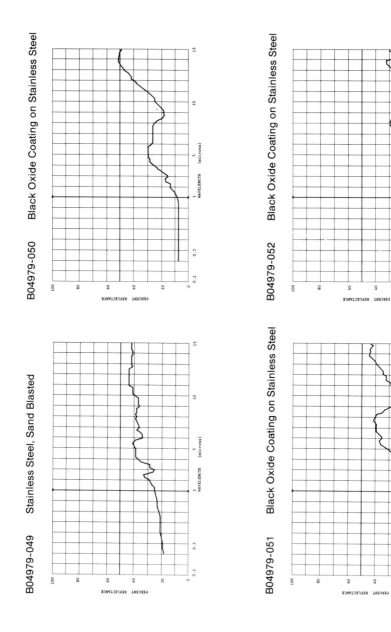

237

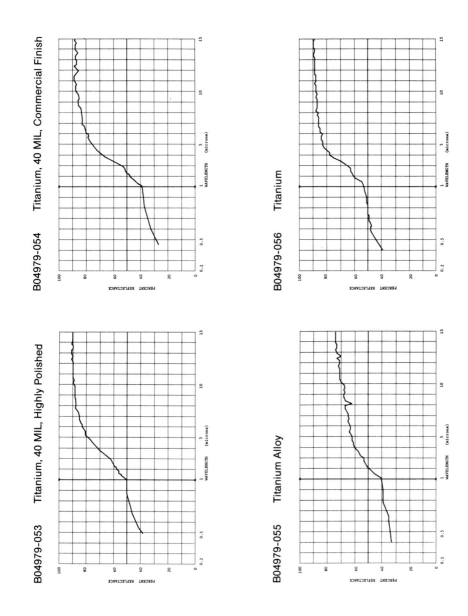

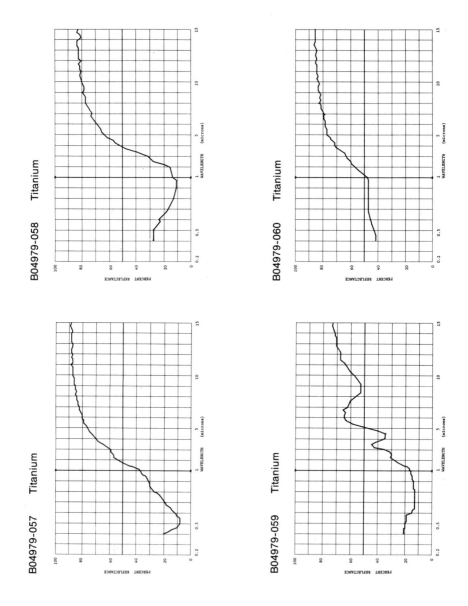

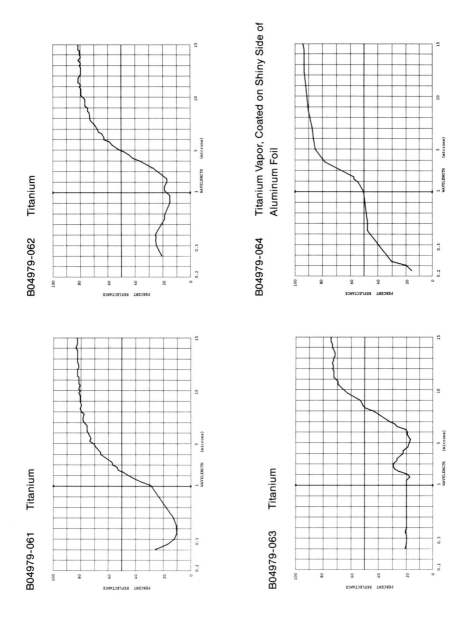

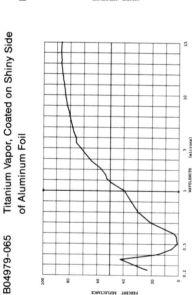

B04979-065 Titanium Vapor, Coated on Shiny Side of Aluminum Foil

B4979-066 Anodized Titanium on Stainless Steel

B04979-067 Polished Zinc on Polished Aluminum

B04979-068 Galvanized Iron, 22 MIL, Commercial Finish

241

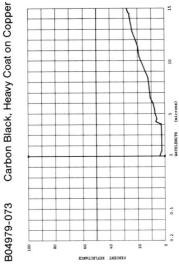

Paint/Metal

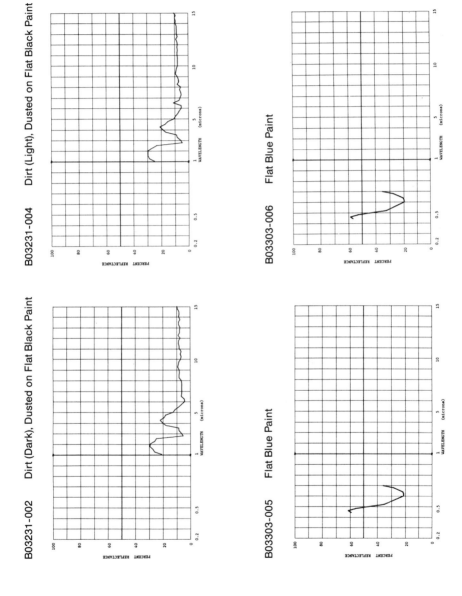

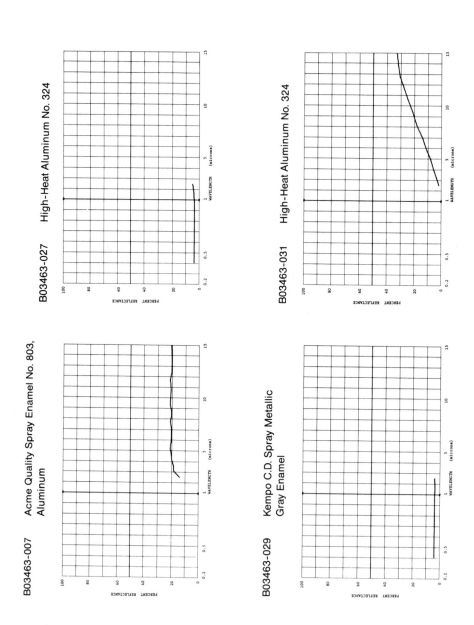

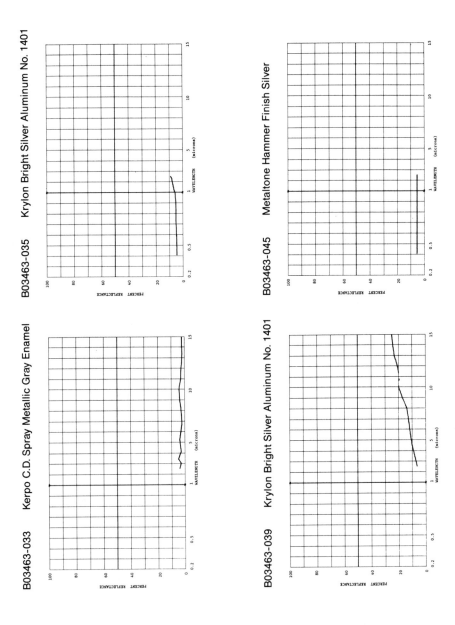

245

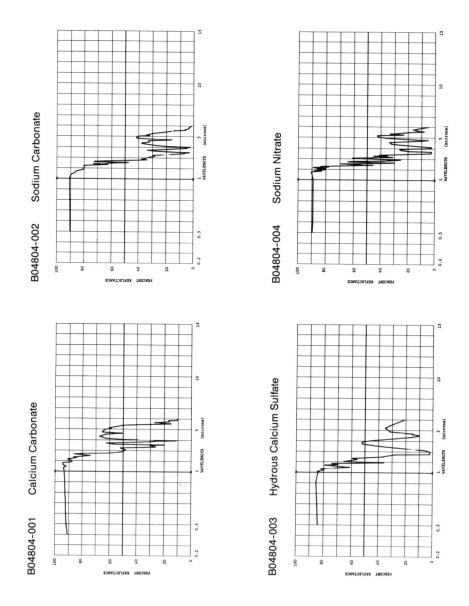

247

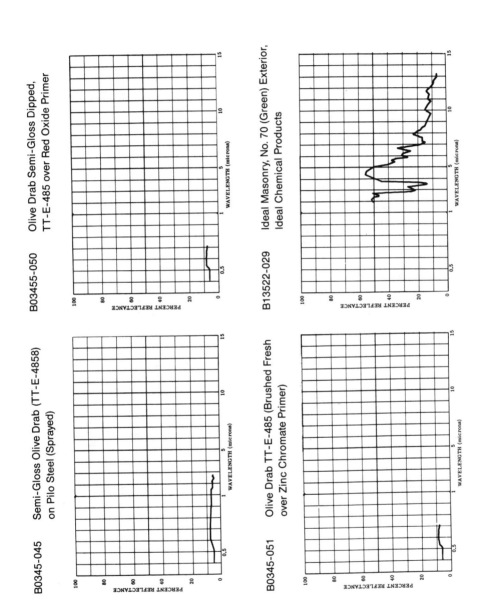

248

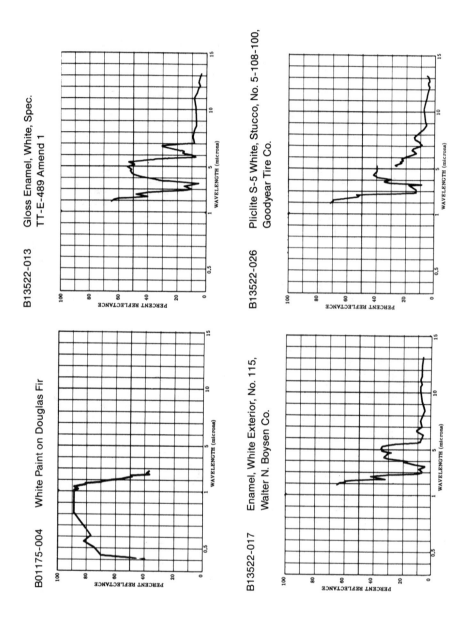

249

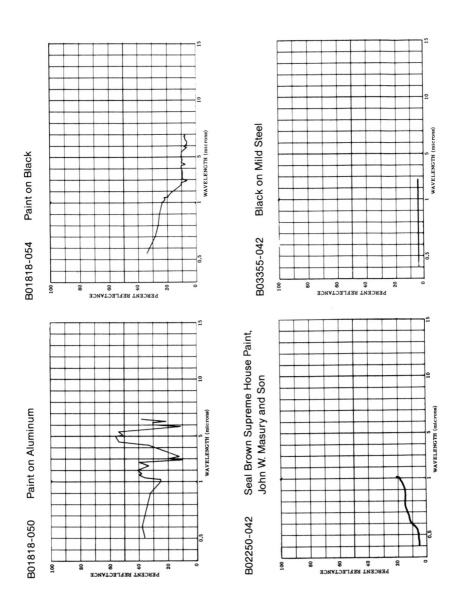

250

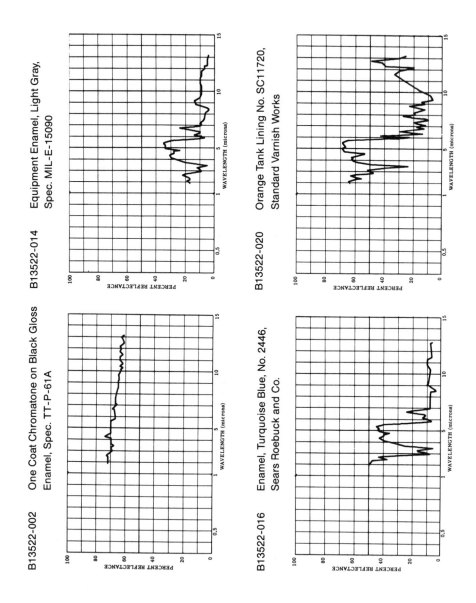

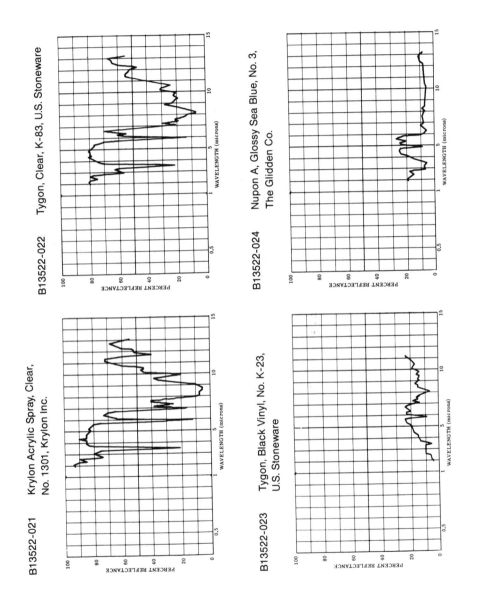

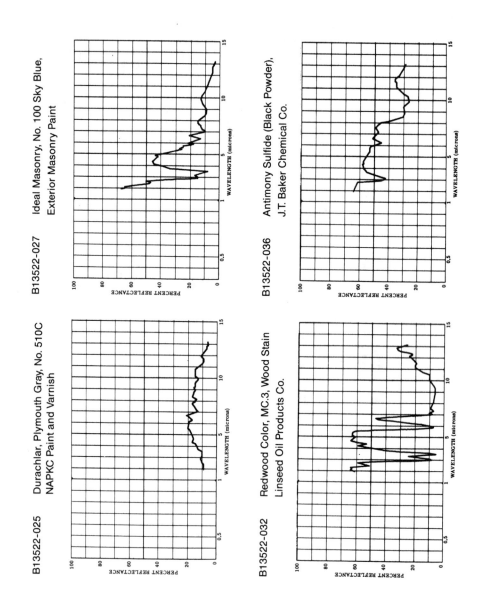

Building Materials

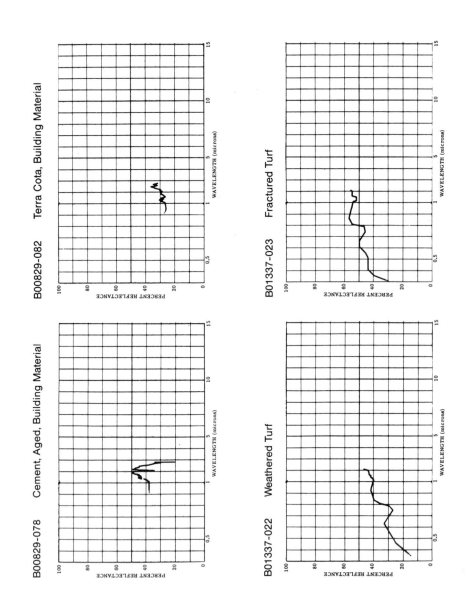

255

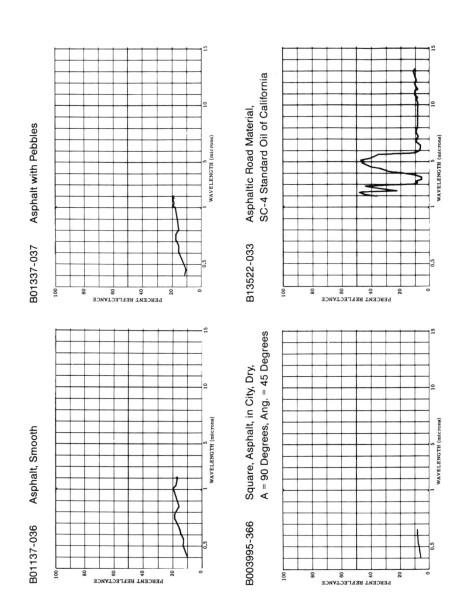

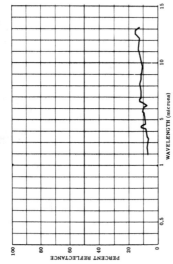

Trees/Leaves

258

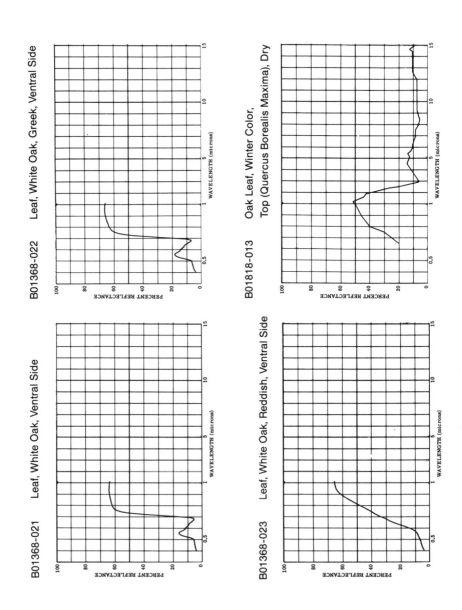

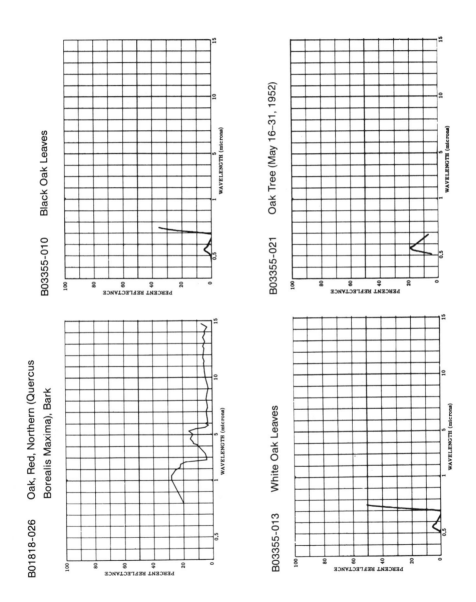

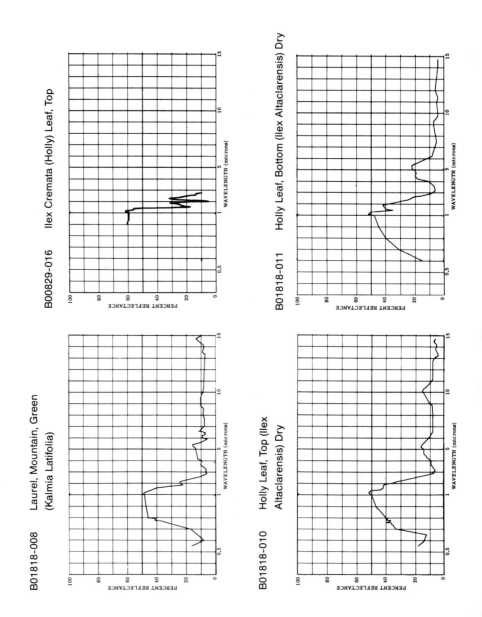

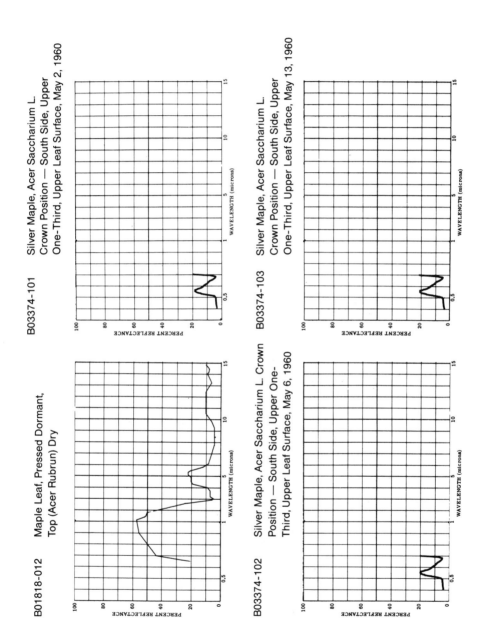

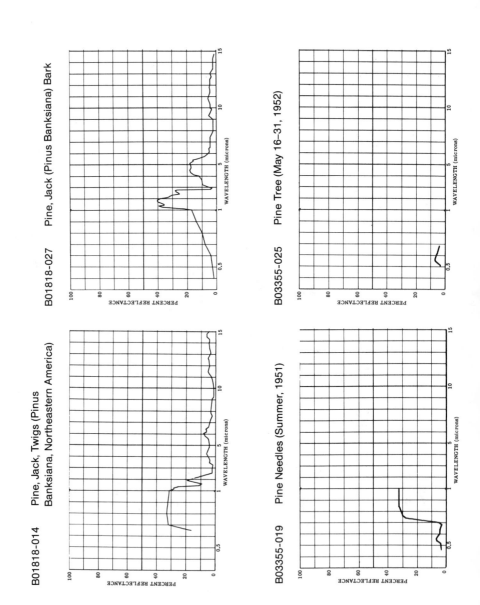

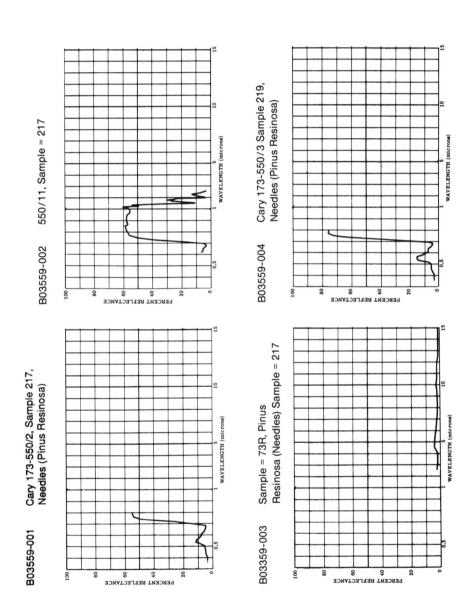

264

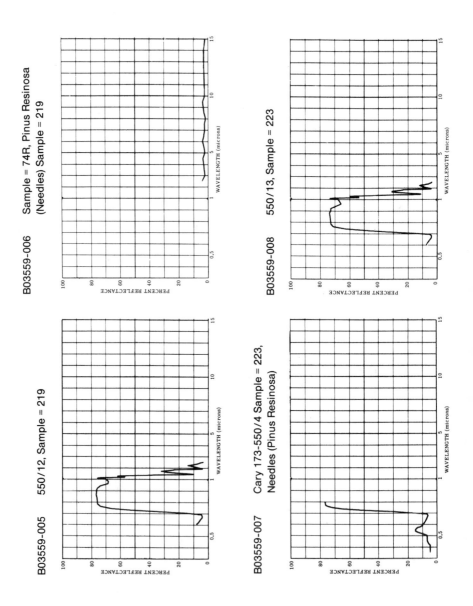

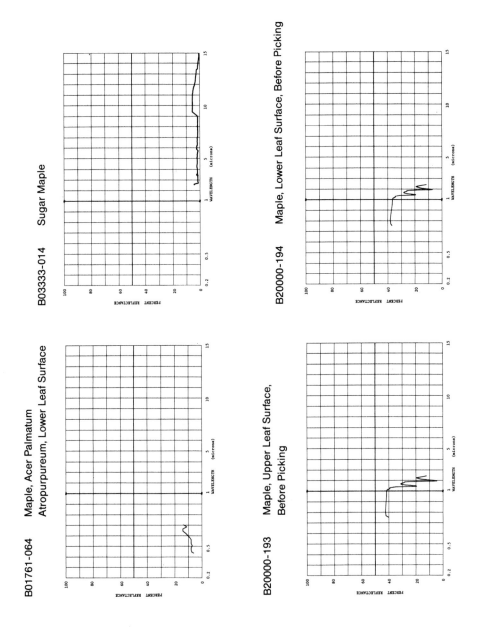

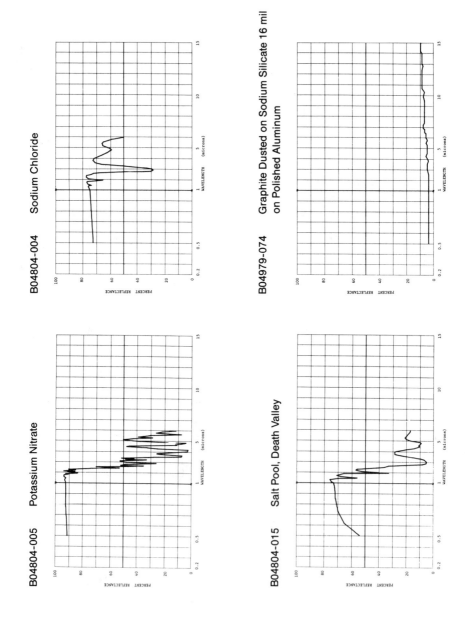

B04979-077 Fluride Dusted on Sodium Silicate 12 mil on Polished Aluminum

B04979-082 Mica, 1.5 mil on Polished Aluminum

B04979-085 Quart Dusted on Sodium Silicate, 30 mil on Polished Aluminum

B04979-086 Pyrite Dusted on Sodium Silicate on Polished Aluminum

268

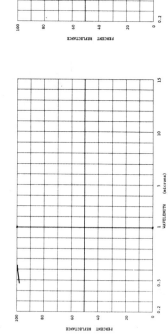
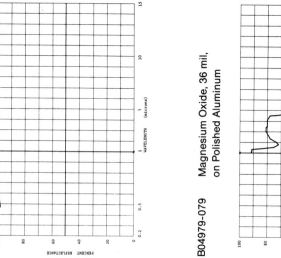
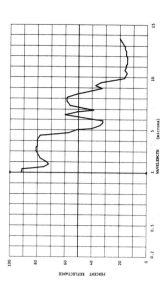

B03257-001 Magnesium Oxide Coefficient of Reflection

B04979-078 Magnesium Oxide, 10 mil, on Polished Aluminum

B04979-079 Magnesium Oxide, 36 mil, on Polished Aluminum

B04979-080 Magnesium Oxide, 40 mil, on Black Base

Soils

270

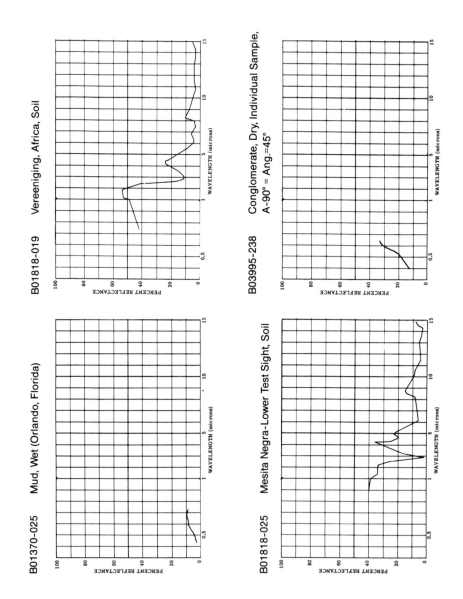

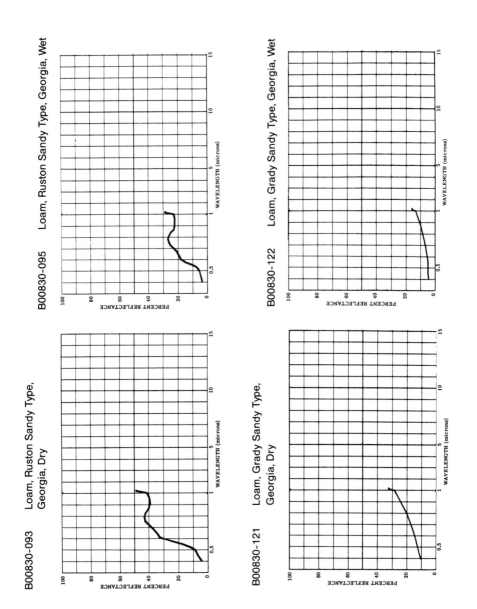

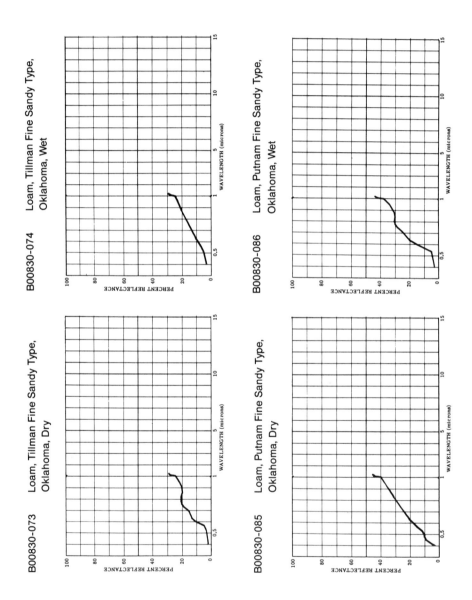

273

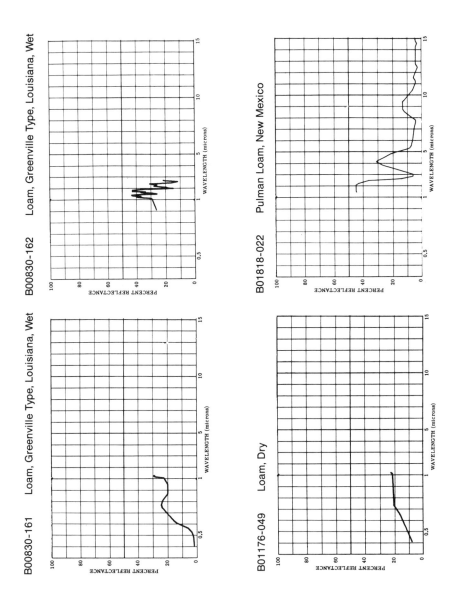

274

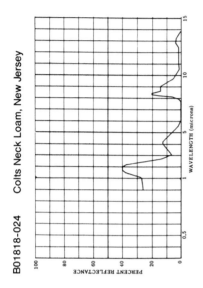

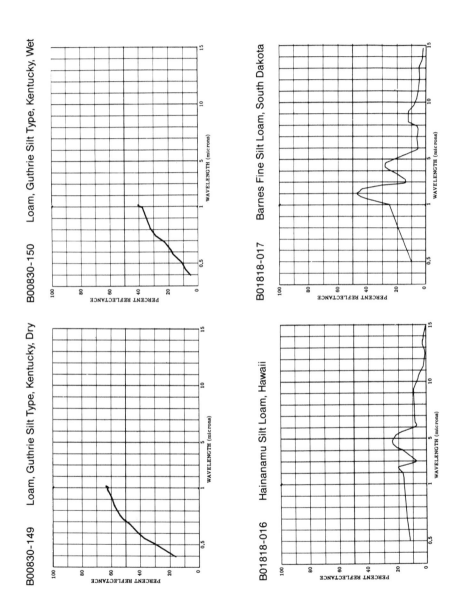

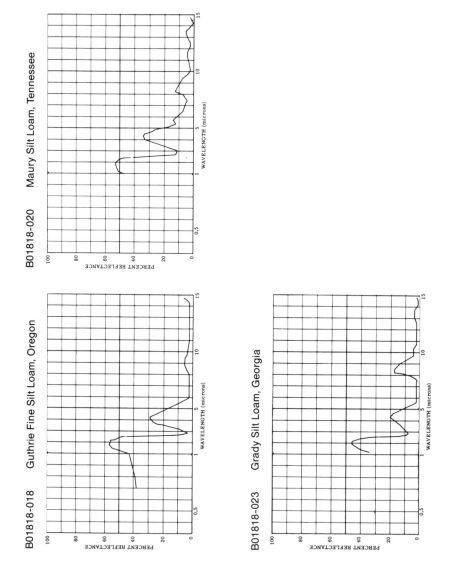

B01818-021   Dublin Clay Loam, California

B03995-311   Soil, Clay Loam, Ploughed, Moist, from the Air, Alt. = 300M

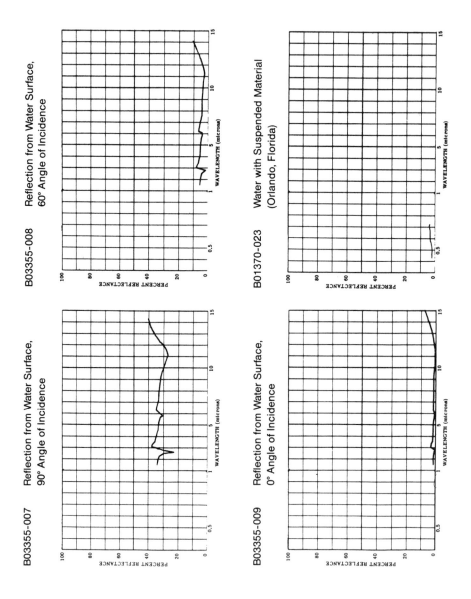

Glossary

Å	angstroms (10,000 micrometers)
A_A	aperture area (M^2)
A_C	receiver collection area
A_{MAX}	target maximum area (m^2)
A_R	cross sectional area of reflecting target
A_S	area to be searched (meter^2)
A_d	detector area (cm^2)
$B_A(z)$	aerosol scattering coefficient (m^{-1} ster^{-1})
B_e	effective bandwidth (Hz)
$B_m(z)$	molecular scattering coefficient (m^{-1} ster^{-1})
$B\pi$	atmospheric backscatter coefficient (m^{-1} ster^{-1})
C, c	speed of light $\approx 3 \times 10^8$ M/S
C_N^2	atmospheric phase structure function (cm$^{-2/3}$)
CNR	carrier to noise ratio
CNR_s	carrier to noise ratio speckle target (diffuse)
CNR_g	carrier to noise ratio glint target
D	aperture diameter (meters)
D_{eff}	effective optics diameter (cm)
D^*	specific detectivity (cm $-$ Hz$^{1/2}$/W)
E_R, E_{sig}, E	received target energy (joules)
E_T	transmitted energy (joules)
F	focal length of detector focusing lens (meters)
F_{AM}	AM modulation frequency (Hz)
F.T.	frame time (seconds)
G	detector gain
G_T	aperture gain of transmitter
I_{GNR}	incoherent Gaussian noise limited receiver
I_1	square of the amplitude
I_{DK}	detector dark current (amp)

I_e	radiant emittance
I_0	the average of the amplitude squared
I_S	atmospheric scatter coefficient
$I_{\lambda d\lambda}$	flux radiation (watts/sq centimeters)
°K	degrees Kelvin
K_V	wave vector (m^{-1})
K_a	aperture illumination function
$\bar{K}$	average number of photo events
K_1	fraction of solar radiation passing through the atmosphere
L	atmospheric path length (meters, kilometers), range
L_1, L_2	near and far range of target
M	number of speckles
M_p	number of photons
N, I	number of cells, or samples
$\bar{N}_N, \bar{n}_N$	average number of noise photoelectrons emitted per unit time
N_R	number of receiving photons
$\bar{N}_S$	average number of signal photoelectrons per spatial correlation cell
N_T	number of transmitter photons per second
N_o	noise power per cycle (W/Hz)
NEP	noise equivalent power (w)
NF	noise figure of receiver
$N(z)$	number density at an altitude z
P_{BBT}	total Blackbody radiation power
P_{BK}, P_{BB}	Blackbody background power (watts)
P_{CW}	continuous wave power
P_D	probability of detection
P_D	probability density
P_{DK}	equivalent dark current power
P_F, P_{FA}, P_N	probability of false alarm
P_{LO}	local oscillator power (watts)
P_{NS}	atmospheric solar scatter power (w)
P_{PK}	pulse peak power (watts)
P_R, P_{sig}	received signal power (watts)
P_{S+N}	probability of r signal plus noise photoelectrons
P_{SB}	solar backscatter power (watts)
P_T	transmitter power (watts)
P_{TH}	equivalent receiver thermal noise power
R	range to target (meters)
R_R	range from receiver to target (m)
R_0	normalized range
G_A	Gaussian beam radius

$R_{\text{CROSSRANGE}} R_C$	cross range resolution (m)
$R_{\text{DOWNRANGE}} R_C$	down range resolution (m)
R_L	detector load resistance (ohms)
R_T	target radius (M)
R_{UNAMB}	range ambiguity (meters)
S	mean signal power (w)
S_{AT}	size distribution parameter
S_{IRR}	solar irradiance (watts/M^2-μm)
S_T	complex aperture obscuration function
S_a	diffuse signal power
SNR S/N	signal to noise ratio
SNR_O	signal to noise ratio (optical)
SNR_m	signal to noise ratio (microwave)
T	temperature (°K)
T_B, TBW	time bandwidth product
T_{FM}	chirped wave form pulse length (sec)
T_P	pulse burst length (sec)
T_s	interpulse spacing (sec)
T_S	system noise temperature
T_{SM}	discrete scatterer smearing time (sec)
T_V	threshold value
T_a	atmospheric transmission
T_e	effective time duration (sec)
T_o	observation time
T.O.T.	time on target (seconds) or measurement interval
V	target velocity (m/sec)
V_{RES}	velocity resolution (m/s)
V_{UNAMB}	velocity ambiguity (m/s)
V_1	visibility
W	water vapor mixing ratio
W_A	aerosol mixing ratio
W_p	average photon counting rate
W_s, $\dot{\theta}$	angular scan rate (rad/sec)
Wg	average photon counting rate due to glint target
Ws	average photon counting rate due to speckle target
$w(x,\bar{x})$	probability density for input S/N ratio
z	altitude (kilometers, meters)
a	amplitude
a_o	average amplitude
d	wire or linear target width (M)
dA	target area (M^2)

f	focal length equal to range in equation 7.8
f	frequency (Hz)
f_d	doppler frequency (Hz)
f_{dmax}	maximum doppler frequency (Hz)
h	Planck's constant (6.626×10^{-34} joule-seconds)
hf	energy per photon (joules)
i_{TH}^2	mean square thermal noise current
i_{BK}^2	mean square background noise current
i_{DK}^2	mean square dark noise current
i_{GR}^2	mean square generation—recombination noise current
i_{LO}^2	mean square local oscillator noise current
i_N^2	mean square noise current
i_{SN}^2	mean square shot noise current
i_{sig}^2	mean square signal current
k	Boltzmann's constant (1.39×10^{-23} J/°K)
k_{MW}	absorption coefficient (Km^{-1}) mid latitude winter
k_{SW}	absorption coefficient (Km^{-1}) sub arctic winter
k_{TROP}	absorption coefficient (Km^{-1}) tropical atmosphere
m	modulation index
n	the number of photo events
n_f	false alarm number (number of samples/p_{FA})
$\bar{n}_S$	average number of signal photoelectrons per unit time
n_T	photoelectron threshold value
$p(I_1)$	intensity probability density function
$p(a)$	amplitude probability density function
$p(n, \tau)$	probability of counting events in time τ
$p(v)$	Rician probability density function
$p(a)$	log normal probability density function
$p(y)$	probability density function of log of the intensity
q	electron charge (1.602×10^{-19} coulombs)
r	radius
x	signal amplitude
x	input signal to noise power ratio
$\bar{x}$	average of S/N over all target fluctuation
y	signal intensity plus noise intensity
z, z'	z axis shifted to z axis
ΔF	source bandwidth (Hz)
ΔF_{SB}	scan spectral bandwidth (Hz)
$\Delta L, \Delta R$	spatial range resolution wherein 50% of an atmosphere signal is received

ΔR_C	coherence distance (m)
$\Delta f, b, B$	electronic bandwidth (Hz)
ΔT	coherence time (seconds)
ΔT	interpulse time
$\Delta\lambda \; d\lambda$	optical bandwidth (micrometers)
Φ	angle between target spin vector and position on a plane aspect angle (degrees)
Ω	scattering steradian angle of target (steradians)
Ω_R	solid angle of received beam (steradians)
Ω_R	solid angle over which energy radiates from radiating body (steradians)
Ω_S	solid angle to searched (steradians)
Ω_T	solid angle of transmitted beam (steradians)
α	particle size parameter
α_S	attenuation coefficient (Km^{-1})
α_1	$\sqrt{\dfrac{2S}{N}}$
ε	target emissivity
η_{ATM}	atmospheric transmission factor
η_{Het}	heterodyne system efficiency
η_{SYS}	system transmission factor
η_d	detector quantum efficiency
θ_B	beamwidth (radians)
θ_T	transmitter beamwidth (radians)
θ_c	angle through which object turns in processing time (degrees)
λ	wavelength (micrometers)
ρ_T	target reflectivity
ρ_i	current responsivity (amps/watt)
ρ_1	ratio of mean to median value of the signal amplitude
τ	pulse length (sec) time interval
ω	target angular rotation rate (Rad/sec)
$\psi(f)$	noise power spectral density (watts/Hertz)
σ	effective target cross section (M^2)
σ_{CR}	corner reflector cross section (M^2)
σ_H	beam misalignment
σ_T	Stefan Boltzmann constant 5.67×10^{-12} W CM^{-2} K^{-4}
σ_{Te}	one sigma estimate of time (sec)
σ_{ext}	extended target cross section (M^2)
σ_f	one sigma estimate of frequency (Hz)
σ_m	median of the signal amplitude

σ_θ	one sigma estimate of angle error
σ_r	one sigma estimate of range (meters)
σ_v	volumetric effective cross section
σ_{wire}	linear (wire type) cross section (M^2)
σ_x^2	variance due to turbulence
σ_x^2, σ_s^2	variance due to turbulence induced scintillation on an aperture
ν	wave number (cm^{-1})

INDEX

Absorption, incoherent measurement approach, 184
Accuracy, of radar measurements, 45
Active/passive imagery
 hanger example, 164
 MTI detection, 163
 simultaneous, 162
Aerosol components
 atmospheric laser radar and, 179–180
 Doppler backscattering and, 192
 remote measurement of, 194
Aerosol mixing ratio, comparison, 188, 190
Aimpoint selection, 164–165
Airborne moving target indication (AMTI), 151, 153
 detection of M-48 tank example, 156
Airborne moving target indication (AMTI) *See also* Moving target indication (MTI)
Airborne pulse Doppler laser radar, 197–198, 201
 A-scope display of, 198, 203
 hardware specifications, 202
Aircraft
 range imaging of, 160, 164
 trailing vortex-time sequence display, 195–197
Alignment
 coherent system, requirements, 36–37
 spatial coherence, 36
All-weather long-range system operation, 24
Amplitude-modulation (AM) CW laser received signal, 154, 159
Angle-angle imaging, 48–49
 floodlight field, 49
 resolved beam, 48
Angle measurements, 43, 46
Angular resolution limitations, 24
Angular scan rate, 37–38
AN/GVS-5 hand-held laser rangefinder, 137, 141
AN/TVQ-2 GLLD, 137–138, 142
Aperture
 diameter, for coherent receivers, 38–39
 illumination function, 24
Argon laser, 121
 Firefly II experiment and, 178
 Firepond facility and, 175, 176
A-scope display, 198, 203
 velocity range (RVI) and, 200, 203
Atmosphere, system efficiency and, 33
Atmospheric
 attenuation, 61
  coefficients for horizontal transmission at sea level, 62
 intensity/time delay, 186
 laser radar systems, 179–206
 propagation, 59–75
 solar scatter equation, 15
 transmission, 60
  computer models and, 61
 turbulence
  absence of in microwave radar model, 89
  effects on laser radar model, 91–96
Atom hill, 122
Attenuation, 59–75
 atmospheric, 61
 coefficients, 61
  for laser frequencies, 66–71
  variation, 63
  versus rain date, 73
 as function of wavelength, 59–61
 microwave radar models and, 72
 radiation fog and, 61
 rain, 61
 in snow, 74
 versus wavelength, 64–65
Autonomous guidance for tactical stand-off weapons, 157
Azimuth measurement, 2
 VAD and, 205

Background noise
 limited incoherent receiver, 21
 mean squared, term, 17
 terms, 13–16
Backscatter
 measurements, near field CW Doppler experiments, 196–197, 199
 properties, atmospheric laser radar, 179–183
Backscattering
 coefficients, 179–180, 186, 200
 function model, 181
 product of, function, 182
Beamshape, system efficiency and, 33
Beam wander, atmospheric turbulence and, 91
Beamwidth, 23–26
 aperture illumination and, 24
 lasers and, 26
 microwave radar systems and, 25
Bidirectional reflectance, 215–217
Bistatic systems
 backscatter coefficient and, 197, 200
 near-field operation of, 192
Blackbody radiation, 10
 equation, 13
 radiant emittance and, 15–16
 spectral radiant emittance and, 11
Bose-Einstein statistics, for radar and lidar

285

signal, 117–118
Brandeiwie, R.A., 197
Bridge detection, classification, and aimpoint selection, 160, 165
Burst parameters, 54
Chemical lasers, 121
Chromium aluminum laser, 121, 122
CNR (carrier-to-noise ratio), 89
CO_2 laser radar
  capabilities, 160
  fire control parameters, 166
  Firefly II experiment and, 178
  flight test, 196, 200
  map, 155
  range images, 157, 159
CO_2 laser radar *See also* Laser radar
Coefficients, attenuation, 61
  for laser frequencies, 66–71
  variation, 63
  versus rain date, 73
Coherence, system efficiency and, 33
Coherent CO_2 imagery, 155
Coherent detection receiver, 13, 14
  effective aperture diameter for, 38–39
  incoherent receiver comparison, 19–22
  local oscillator signal and, 17
  system organization and, 30
Coherent detection system, 35
  beam wander and scintillation effects on, 91
  for fire control application, 166
  laser radar model, 119
  phase front distortion and, 34
  receiver aperture area and, 38
  statistics, 77
  temporal and spatial coherent requirements and, 34
Coherent laser radar
  block diagram, 152
  figure of merit for, 27–28
  models, 88–96
    atmospheric turbulence effects on, 91–96
  systems and techniques, 151–178
Coherent optical detection
  receiver
    Geos-III tracking and, 169, 172
    noise figure, 23
  system
    detection probabilities, 77
    photomixing spatial requirements for, 35
Coherent systems, atmospheric, 192–206
  far field operation and, 193–195
  far-field pulse Doppler measurements and, 197–206
  near field CW Doppler experiments and, 195–197
  near-field operation and, 192–193
Color scale bar, 154
Copperhead cannon-launched guided projectile (CLGP), 138, 143
Correlation cells, in energy-detection radar, 100
Cross-range resolution, 51–52
Cross-range velocity, 47–48
Cross section
  imaging, 8
  Lambertian diffuse point target, 8
  triangular corner reflector, 5
  wire target, 7
CW-based waveforms, 43–44
Data rate transmitters, laser radar, 26
Davis, W.C., 197
Deirmendjian atmospheric model, 180
Depolarization for targets, 32
Detection, 10–23
  background noise terms and, 13–16
  box, 160
  incoherent and coherent receivers and, 19–22
  noise figure and, 22–23
  noise power spectral density and, 12
  noise receiver and signal terms and, 17–19
  receiver, techniques, 13
  SNR expression development and, 16–17
  statistics for glint, speckle, and semirough targets, 96
  statistics (microwave radar models), 78–87
Detection probability
  comparison, 82
  glint target, no-turbulence, 88–89
  glint target, turbulence, 92
  for incoherent Gaussian noise receiver, 108
  incoherent systems, 97–108
  speckle target, no-turbulence, 89
  speckle target, turbulence, 95
Detector, 131–136
  characteristics, 135–136
  current, 131, 132
  responsivity, 132–133
  dark current, 17
  equation, 18
  incoherent receiver and, 30
  measurements, 135
  operation in square law region, 131–132
  quantum efficiency, 23
  responsivity, 19, 132–133
  selection for incoherent receiver, 29–30
  system efficiency and, 31
  typical, sensitivities and characteristics, 133–134
Deterministic field: Poisson distribution, 117–118
Deterministic field plus narrowband Gaussian field, 118
Differential absorption, incoherent measurement approach, 184
Diffraction-limited beamwidth, 23
Diffuse target
  range equation for, 6
  target coherence time for, 52
Diffuse target *See also* Speckle target
DiMarzio, C.A., 103
Diode-laser-pumped zig-zag slab laser
  average power output from, 129
  concept, 127
  thermal controlling of, 131
Direct detection, incoherent measurement approach, 184
Directional reflectance, 217–218

Discrete scatterers, coherent observation time
    limitations, 52
Doppler
    beam shaping techniques, 49
    broadening, 195
    and range ambiguities, 55
    shift
        equation for, backscatter signal, 49
        far field operation and, 193–195
        PRF sampling and, 50–51
        produced by moving targets, 50
        signal illustration, 151, 153
        space objects spectra results, 175
Down-range resolution, 51–52
Edge extraction techniques, 154, 159
Efficiency, system, 30–33
Electromagnetic scattering theory, 215
Electromagnetic spectrum, 1–3
    beamwidths and, 24
Electronic receiver/processor loss, system
    efficiency and, 31–33
Elevation angle measurement, 2
Energy-detection radar, 100
    non-fluctuating signal and, 97
    rough target, performance, 102–103
    specular target, performance, 101
Equivalent inverse synthetic aperture radar
    (ISAR), 49
Extended target, range equation for, 6, 7
False alarm indications, MTI detection, 153
False alarm probability
    glint target, no-turbulence, 88
    glint target, turbulence, 92
    incoherent systems, 97–108
    speckle target, no-turbulence, 89
    speckle target, turbulence, 95
Fante, 91
Far field operation
    coherent systems and, 193–195
    pulse Doppler measurements and, 197–206
Far field, range equation in, 4
Far-infrared wavebands, 1, 2
    atmospheric propagation and, 59–61
    detector characteristics and, 135–136
Fire control laser radar adjuncts, 166–178
Firefly II laser radar experiment, 177–178
Flight tests, 161
Floodlight field imaging, 49
Flourescence, incoherent measurement
    approach, 184
Fluctuation models, 78–79
    comparison, 81
FM systems, measurement errors and, 45–46
Focused systems, near-field operation of, 192
Forward-looking ground based systems, 154
Frequency-modulated waveforms, 43–44
Gallium arsenide lasers, 121, 125
    diode lasers with integral SI heat sink and,
        126
Gaussian beam, aperture illumination with, 24
Gaussian noise receiver, target detection and, 78
Generation recombination noise
    equation, 18
    SNR and, 17
Geos-III satellite
    elevation angle tracking statistics of, 173
    tracking of, 169–175
Glint target
    detection statistics, 96
    no turbulence-detection and false alarm
        probability, 88–89
    pulse integration, 93
    radar and lidar signal statistics for, 117–118
    Rician distribution and, 115
    turbulence-detection and false alarm
        probability, 91–92
Goodman, J.W., 100
Green light, 121–122
Ground laser locator designator (GLLD)
    AN/TVQ-2, 137–138, 142
    CLGP and, 138
Harney, R.C., 91
Haystack X-band imaging, in Firefly II
    experiment, 177
Haze
    atmospheric laser radar and, 180
    YAG system range performance in, 169
Helicopter targets, 154, 158
Helium neon laser, 121
Heterodyne
    efficiency
        coherent systems and, 36–37
        versus angular alignment requirements,
            37
    system configuration, 204
Horrigan, F.A., 192, 194
IGNR
    incoherent/coherent SNR comparison,
        single sample SNR for, 108
    one hundred aperture averaged for, 106
    single speckle diffuse target for, 104
    ten samples averaged times ten
        measurements averaged for, 107
    ten speckles aperture averaged for, 105
Imaging, 8
    simultaneous active/passive illustration,
        162–164
    systems, 48–57
        angle-angle imaging, 48–49
        range-doppler imaging, 49–51
        range-velocity ambiguity and, 51–57
Incoherent detection
    laser radar model, 119
    of optically noise (Poisson) limited receivers,
        100–109
    YAG laser radar system, 147, 149
        configuration and specifications, 147–148
Incoherent detection receiver, 13, 14
    coherent receiver comparison, 19–22
    non-fluctuating signal detection and, 97
    system organization, 29–30
    systems and techniques, 137–150
    target smoothness characteristics and, 100
*Incoherent Optical Signal Detection in Gaussian
    Noise,* 103

Incoherent radiation sources
  bandwidth, 33
  spectral content, 40
Incoherent systems, atmospheric measurement, 184–191
  approaches, 184
Infrared systems, adverse weather capability of, 24
Integrated improvement factor, 84
Intensity/time delay, atmospheric, 186
Intensity-velocity (IVI) display, 200–201, 205
Introduction to Radar Systems, 82–83
Isotropic target, range equation, 6

Junge atmospheric model, 180

Kachelmyer, A.L., 55

Ladar (laser detection and ranging), 1
Lag angle scan, 37–38
Laguerre statistics, for radar and lidar signal, 117–118
Lambertian
  diffuse point target, 8
  target, 6
  wire target, 7
Laser
  altimeter system, 137
    Apollo, 139
    equipment specifications, 139–140
    hardware for Apollo, 140
  detection statistics, 97–99
    distinctions between microwave and, 109
  linewidth measurements, 40–41
  range equation (radar range equation), 9
  wavebands, atmospheric propagation and, 59
  wavelengths, 121
    detection and, 10
    Doppler shifts and, 46
Laser Doppler velocimeter (LDV) van, wind shear measurement and, 205, 206
Laser radar
  atmospheric, systems, 179–206
  capabilities, 157, 160
  coherent detection models, 119
  in conjunction with microwave and passive IR systems, 166
  incoherent detection models, 119
  limited data rate, 166
  long-range, 169–178
  pulse Doppler performance calculations, 167
Laser radar *See also* CO_2 laser radar
Laser rangefinders
  10.6-μm, 147, 150
    parameters, 150
  AN/GVS-5 hand-held, 137, 141
  calculated peak laser power for four, 147
  functional block diagram of, 137, 138
  parameters and performance calculations for four, 146
Laser-reflected energy, 138, 145
Laser Remote Sensing, 184
Lasers, 121–136
  basic elements of, 121–122
  coherent operation of, 33–34
  detectors and, 131–136
  popular sources for, 129–130
  types of, 121
  user's guide to, 131
Lawrence, M.P., 179, 184
Lidar (light detection and ranging), 1
  coherent detection systems and, 192
  detection probabilities and, 77–78
  signal statistics in the photon count limit, 117–119
Limited data rate laser radars, 166
Lincoln Firepond 10.6 um laser Doppler radar
  1980 tracking results, 174
  Doppler spectra results from, 175
  facility, 170
  improved facility, 175, 176
  satellite tracking mode of, 172
  schematic, 171
  specifications, 170
  uses of, 169–178
Linear frequency modulated chirp waveform, 55–57
Linear propagation, to space, 75
Linear target, 6, 7
Linewidth measurements, laser, 40–41
Load oscillator induced noise, 18
Local oscillator power
  coherent detection and, 20
  spatial coherence alignment and, 36
Log-normal distribution, 114–115
  amplitude, 115
  function, 114
Long-range laser radar, 169–178

McCormick, M.P., 179, 180, 184
Marchege, J.F., 43
Marcum, J.I., 77, 78
Master oscillator power amplifier (MOPA), 197, 201
Measurement errors, 42–48
Measures, Raymond, 184
Melfi, S.H., 184
Microwave radar systems, 2–3
  in conjuction with laser radar systems, 166
  figures of merit for, 27
  multipath errors and, 25–26
  target detection statistics, 78–87
Microwave receiver
  noise figure, 23
  SNR power for, 22
Microwave wavebands, 1
  atmospheric propagation and, 59
  corner reflectors at, 5
  detection and, 10
  Doppler shift and, 50–51
Mid-infrared waveband, 1, 2
  atmospheric propagation and, 59
  detector characteristics and, 135–136
Millimeter waveband, 1, 2
  and adverse weather capability, 24
Millstone Hill radar, 177
  Firefly II radar experiment and, 177
  GEOS-III satellite and, 169–170

Mirror type target, range equation for, 5–6
MMIC (monolithic microwave integrated circuits), 125
Molecular constituent excitement, 188, 191
Moving target indication (MTI)
  active/passive images and, 163
  evaluated for Raytheon air-cooled 5W laser, 152–154
  target detection requirements, 153
Moving target indication (MTI) *See also* Airborne moving target indication (AMTI)
Multipath
  errors, 25
  lasers and, 26
  limitations, 24

Narrowband optical Gaussian field: Bose Einstein distribution, 118
NASA 90 pod configuration, 198, 202
Near-field operation
  coherent systems and, 192–193
  CW Doppler experiments and, 195–197
Near field, range equation in, 4–5
Near infrared waveband, 1, 2
  detector characteristics and, 135–136
Negative binomial statistics
  for radar and lidar signal, 117–118
  rough target and, 100
Neodymium lasers, 121
Neodymium YAG laser, 123, 125
  absorption spectrum, 127, 128, 131
  adjunct system parameters for fire control, 168
  diode-laser-pumped zig-zag slab laser and, 127, 128, 129
  Firepond facility and, 175, 176
  miniaturized pulsed, transmitter, 124, 125–126
  ruby system comparison, 144, 146
  sensitivity of, in clear weather, 168
  visibility (haze) sensitivity, 169
Neon laser, 121
Noise equivalent power (NEP), 29–30
  detectors and, 133–136
Noise figure, of optical receiver, 22–23
Noise power spectral density, 12
  versus wavelength, 12
Non-fluctuating model, 80
  comparison with cases, 1, 2, 3, 4, 81
Optical
  beamwidths, 23–24
  detector, 132
    coherent detection and, 13
    responsivity of, 132–133
    heterodyning, 36
    range equation, 8
    receivers, 13
Optically noise (Poisson) limited receivers, 100–108
Optical Studies of the Atmosphere, 179
Optics, system efficiency and, 30

Palmer scan geometry, RPV, 154

Paport, R.L., 91
Passive/active imagery
  hanger example, 164
  MTI detection, 160, 163
  simultaneous, 162
Passive IR systems, 157
  in conjuction with laser radar systems, 166
Passive sensors, with laser radar systems, 3
PATS incoherent detection YAG Laser radar, 148
Pave Penny pod, 138, 144
Photoconductor detector, noise term for, 18
Photomixing spatial coherent detection, 34, 35
Photomultiplier systems, 144, 146–147
Photon energy, versus wavelength, 11
Pointing resolution, from laser radar quadrant-photodiode detector, 46–47
Point target
  Lambertian diffuse cross section, 8
  range equation for, 6, 7
Poisson statistics
  non-fluctuating signal detection and, 97
  for radar and lidar signal, 117
  specular target detection and, 100
Polarization-controlled optical modulator, 197–198, 201
Polarization, system efficiency and, 31
Popular inversion, 122
PRF sampling, Doppler shift and, 50–51
Probability density function, 80
Probability of detection
  incoherent systems, 97–108
  versus additional SNR, 83
  versus probability of false alarm, 85–87
  versus SNR with neodymium laser, 166, 168–169
  versus SNR per pulse, 82
Probability of false alarm
  incoherent systems, 97–108
  versus probability of detection, 85–87
Propagation
  atmospheric, 59–75
  linear, to space, 75
  to space, 74, 75
Pulse
  compression techniques, 43–44
  integration
    glint target, 93
    speckle target, 94
  transmitter, for range resolution, 193
  waveforms, 43–44, 56
Pulsed flashpump spectrum, 127, 128
Pulse Doppler laser radar performance calculations, 167
Pulse-to-pulse fluctuation, 79
  applications, 80

Quantum efficiency, 23
  incoherent receiver and, 29
Quantum noise-limited coherent optical receiver, 22–23
Quantum noise (Poisson), 13

Radar

measurement accuracy, 45
signal statistics, 117
Radar *See also specific types of radar*
Radiant emittance, blackbody, 15–16
Rain attenuation, 61
Raman scattering
   incoherent measurement approach, 184
   remote atmospheric monitoring and, 188
Range ambiguity, 51
   Doppler and, 55
Range-Doppler
   ambiguity function, 55–57
   imaging, 49–51
      Firefly II experiment and, 178
      firepond facility and, 175
Range-Doppler Imaging with a Laser Radar, 55
Range equation, 3–10
   laser, 9
   optical, 8
   target area dependence, 6–10
Range images
   active/passive, 160, 164
   edge extinction and, 159
   laser radar, 158, 159
Range measurement, 2, 46
   accuracy, 44–45, 47
   AM CW laser received signal and, 154, 159
   errors in, 43, 44
   laser radar systems and, 3
Range resolution, for far field operation, 193–194
Rayleigh and Mie scattering theories, 179
Raytheon
   5W air-cooled laser, 151, 154
   5W coherent 10.6 μm airborne system, 151, 153
   A-scope display, pulse Doppler laser radar, 198, 203
   autonomous guidance for tactical stand-off weapons, 157
   pulse Doppler laser radar hardware specifications, 198, 202
   Triservice, system, 154, 157
Receiver
   detection techniques, 13
   systems, 14
Receiver *See also specific types of receivers*
Receiver Johnson noise, 17
Receiver noise, and signal terms, 17–19
Reflectance measurements, 214–218
Reflectivity measurement, 2
Refraction index, turbulence and, 91
Remote atmospheric measurement techniques, 183–206
   coherent systems, 192–206
   incoherent systems, 184–191
   incoherent systems approaches, 184
Resolved beam imaging, 48
Rician distribution, 115–117
   function, 116
   log amplitude, 116
RMS errors, 42, 43
Rod cooling techniques, 124, 125
Rough target, energy-detection radar
   performance, 102–103
Royal Signals and Radar Establishment, 196
RPV Palmer scan geometry, 151, 154
Ruby chromium laser, 121, 122
   atmospheric measurement experimental setup, 185
   Firepond facility and, 175, 176
   incoherent system atmospheric measurements and, 184
   neodymium YAG laser comparison, 144, 146
   Raman scattering and, 188
   relative volume backscatter and, 188
   simplified energy-level diagram of, 123

Scan spectral broadening, 41
Scan-to-scan fluctuation, 79
   applications, 80
Scintillation, atmospheric turbulence and, 91
Search field, 26–27
   equations, 28–29
   figure of merit, 27–29
Seeber, Dr. K., 36–37, 38, 103
Seeker hardware, 160, 161
Semiconductor lasers, 121
Semirough targets, detection statistics, 96
Shapiro, J.H., 91
Signal induced noise, 13
Signal terms, and receiver noise, 17–19
Signal-to-noise ratio. *See* SNR
Silicon detector systems, 144, 146–147
Skolnik, M., 42, 77, 82
SNR
   coherent detection equations, 19, 20
   comparison between incoherent and coherent detection statistics, 108
   detection probabilities, 77
   expression development, 16–17
   far field operation and, 194–195
   of focused system, 192
   for glint target detection/false alarm probability, 88
   incoherent detection equation, 19, 21
   for laser radar models (incoherent and coherent detection), 119
   optimization for coherent receiver, 30
   optimization for incoherent receiver, 29–30
   power
      of microwave receiver, 22
      for quantum noise-limited coherent optical receiver, 22–23
   requirements for target receiver operating characteristics, 90
   for speckle target detection/false alarm probability, 89
   for Swerling IV targets, 114
   versus probability of detection, 82
   versus probability of detection for ND:YAG laser, 168
   versus system range error, 46
   versus transmitter power, 21–22
Solar Backscatter, 15
Sonnenschein, C.M., 192, 194
Spatial coherence, 36
   alignment, 36

of electromagnetic wave, 34
temporal coherence and, 33–41
Spatial effects, turbulence and, 91
Spatial resolution, of near-field operation, 193
Speckles
  10, aperture averaged for IGNR, 105
  10, samples averaged times ten measurements averaged for IGNR, 107
  100, aperture averaged for IGNR, 106
Speckle target
  detection statistics, 96
  no-turbulence, detection and false alarm probability, 89
  photon noise not the dominant receiver term and, 103
  pulse integration, 94
  radar and lidar signal statistics for, 117–118
  single, for IGNR, 104
  system efficiency and, 31
  turbulence, detection and false alarm probability, 95
Speckle target *See also* Diffuse target
Spectral
  content, 40
  detectivities, 134
  irradiance of the sun, 16
  response characteristics for photoemissive devices, 133
Specular target, energy-detection radar performance, 101
Square law region, detector operation in, 131–132
Stefan-Bolzmann law, 10
  radiant emittance and, 15
Steradian search field, 27
Structure function variations, 39
Swerling I, 109
  amplitude distribution, 111
  log amplitude distribution, 112
  power distribution, 113
Swerling II, 109–110
  amplitude distribution, 111
  comparison with Swerling IV, 111–114
  log amplitude distribution, 112
  power distribution, 113
Swerling III
  amplitude distribution, 112
  log amplitude distribution, 112
  power distribution, 113
Swerling IV, 109, 110–111
  amplitude distribution, 112
  comparison with Swerling II, 111–114
  log amplitude distribution, 112
  power distribution, 113
Swerling, P., 77, 78, 88
  four target fluctuation models of, 78–79
Sylvania laser radar, 147
Synthetic aperture radar (SAR), 49
System organization, 29–33
  coherent receiver and, 30
  efficiency and, 30–33
  incoherent receiver and, 29–30

Tank targets, 154, 156–157
Target
  coherent observation time limitations, 52
  detection, 8
    statistics (microwave radar models), 78–87
  fluctuation models, 78–79
    comparison, 81
    integrated improvement factor and, 84
  location determination measurement errors, 42
  measurement with laser radar, 27, 43
  receiver operating characteristics, 90
  reflection characteristics, 213–214
  search function with laser radar, 27
  system efficiency and, 31
Temperature measurement, 2
Temporal coherence, 40–41
  of electromagnetic wave, 34
  spatial coherence and, 33–41
The Infrared Handbook, 61
Thermal noise, 10
  current, 18
  noise power spectral density and, 12
Thermal-receiver noise, 17
Time sequence display
  aircraft trailing vortex, 195–197
  wake vortex, 196, 198
Transmittance
  of combined carbon dioxide and water vapor, 72
  of water vapor, 71
Transmitter power, versus SNR, 21–22
Triangular corner reflector, 5
Truncation loss, system efficiency and, 33
Turbulence. *See* Atmospheric turbulence

Uncertainty principle of radar, 42, 43

Vaughan, J.M., 196
Velocity
  ambiguity, 51
  differential, far field operation and, 193–194
  measurement, 2, 46
    accuracy versus bandwidth, 47
    errors in, 43, 44–45
    laser radar systems and, 3
Velocity azimuth display (VAD), 205–206
Velocity range (RVI) display, 200, 203
Vernier mirror tracking data, for GEOS-III statellite, 175
VHSIC (very high speed integrated circuts), 125
Video radiometer receiver, 13
Volume backscatter
  for aerosols and molecules, 186–188
  cross-section for aerosols and molecules, 188
Volumetric cross section, of atmosphere, 179

Wake vortex-time sequence display, 196, 198
Wallops Island radar, Firefly II experiment and, 175, 177
Water vapor mixing ratio
  comparison, 188, 191
  profiles, 188, 190
Wavebands, 1–2

Waveform
  performance parameters, 54
  pulse and linear frequency modulated chirp, 55–56
  range-Doppler ambiguities and, 57
  range-velocity ambiguity and, 53
  system efficiency and, 33
Weiner, S., 52
Wideband optical Gaussian field: negative binomial distribution, 118
Wind field, in far field operation, 193–194, 195
Wind shear
  atmospheric information, 204–205
  measurement display, 206
Wright, M.L., 180

YAG laser. *See* Neodymium YAG laser

The Artech House Radar Library

David K. Barton, *Series Editor*

Active Radar Electronic Countermeasures by Edward J. Chrzanowski

Adaptive Signal Processing for Radar by Ramon Nitzberg

Airborne Pulsed Doppler Radar by Guy V. Morris

AIRCOVER: Airborne Radar Vertical Coverage Calculation Software and User's Manual by William A. Skillman

Analog Automatic Control Loops in Radar and EW by Richard S. Hughes

Aspects of Modern Radar by Eli Brookner, *et al.*

Aspects of Radar Signal Processing by Bernard Lewis, Frank Kretschmer, and Wesley Shelton

Bistatic Radar by Nicholas J. Willis

Detectability of Spread-Spectrum Signals by Robin A. Dillard and George M. Dillard

Electronic Countermeasure System Design by Richard Wiegand

Electronic Homing Systems by M.V. Maksimov and G.I. Gorgonov

Electronic Intelligence: The Analysis of Radar Signals by Richard G. Wiley

Electronic Intelligence: The Interception of Radar Signals by Richard G. Wiley

Engineer's Refractive Effects Prediction Systems — EREPS by Herbert Hitney

EREPS: Engineer's Refractive Effects Prediction System Software and User's Manual, developed by NOSC

High Resolution Radar by Donald R. Wehner

High Resolution Radar Cross-Section Imaging, Second Edition by Dean L. Mensa

Interference Suppression Techniques for Microwave Antennas and Transmitters by Ernest R. Freeman

Introduction to Electronic Defense Systems by Fillippo Neri

Introduction to Electronic Warfare by D. Curtis Schleher

Introduction to Sensor Systems by S.A. Hovanessian

IONOPROP: Ionospheric Propagation Assessment Software and Documentation by Herbert Hitney

Logarithmic Amplification by Richard Smith Hughes

Modern Radar System Analysis by David K. Barton

Modern Radar System Analysis Software and User's Manual by David K. Barton and William F. Barton

Monopulse Principles and Techniques by Samuel M. Sherman

Monopulse Radar by A.I. Leonov and K.I. Fomichev

MTI and Pulsed Doppler Radar by D. Curtis Schleher

Multifunction Array Radar Design by Dale R. Billetter

Multisensor Data Fusion by Edward L. Waltz and James Llinas

Multiple-Target Tracking with Radar Applications by Samuel S. Blackman

Multitarget-Multisensor Tracking: Advanced Applications, Yaakov Bar-Shalom, ed.

Over-The-Horizon Radar by A.A. Kolosov, *et al.*

Principles and Applications of Millimeter-Wave Radar, Charles E. Brown and Nicholas C. Currie, eds.

Pulse Train Analysis Using Personal Computers by Richard G. Wiley and Michael B. Szymanski

Radar and the Atmosphere by Alfred J. Bogush, Jr.

Radar Anti-Jamming Techniques by M.V. Maksimov, *et al.*

Radar Cross Section Analysis and Control by A.K. Bhattacharyya and D.L. Sengupta

Radar Cross Section by Eugene F. Knott, *et al.*

Radar Detection by J.V. DiFranco and W.L. Rubin

Radar Electronic Countermeasures System Design by Richard J. Wiegand

Radar Evaluation Handbook by David K. Barton, *et al.*

Radar Evaluation Software by David K. Barton and William F. Barton

Radar Propagation at Low Altitudes by M.L. Meeks

Radar Range-Performance Analysis by Lamont V. Blake

Radar Reflectivity Measurement: Techniques and Applications, Nicholas C. Currie, ed.

Radar System Design and Analysis by S.A. Hovanessian

Radar Technology, Eli Brookner, ed.

Receiving Systems Design by Stephen J. Erst

Radar Vulnerability to Jamming by Robert N. Lothes, Michael B. Szymanski, and Richard G. Wiley

RGCALC: Radar Range Detection Software and User's Manual by John E. Fielding and Gary D. Reynolds

SACALC: Signal Analysis Software and User's Guide by William T. Hardy

Secondary Surveillance Radar by Michael C. Stevens

Small-Aperture Radio Direction Finding by Herndon Jenkins

Solid-State Radar Transmitters by Edward D. Ostroff, *et al.*

Space-Based Radar Handbook, Leopold J. Cantafio, ed.

Spaceborne Weather Radar by Robert M. Meneghini and Toshiaki Kozu

Statistical Signal Characterization by Herbert Hirsch

Statistical Theory of Extended Radar Targets by R.V. Ostrovityanov and F.A. Basalov

VCCALC: Vertical Coverage Plotting Software and User's Manual by John E. Fielding and Gary D. Reynolds